BIOLOGY: BRAIN & BEHAVIOUR

Control of Behaviour

Springer
Berlin
Heidelberg
New York
Barcelona
Budapest
Hong Kong
London
Milan
Paris
Santa Clara
Singapore
Tokyo

Frederick Toates (Ed.)

Control of Behaviour

With 77 Figures

Springer in association with

Unless otherwise stated, all contributors are (or were at the time this book was written) members of The Open University

Academic Editor

Frederick Toates

Authors

Stuart Armstrong (La Trobe University, Australia)
Basiro Davey
Frederick Toates
Terry Whatson
Anna Wirz-Justice (Psychiatrische Universitätsklinik Basel, Switzerland)

External Assessors

Richard Andrew, School of Biological Sciences, University of Sussex (Series Assessor)
Trevor Robbins, Department of Experimental Psychology, University of Cambridge (Book Assessor)

***Biology: Brain & Behaviour* series**

1 Behaviour and Evolution
2 Neurobiology
3 The Senses and Communication
4 Development and Flexibility
5 Control of Behaviour
6 Brain: Degeneration, Damage and Disorder

Library of Congress Cataloging-in-Publication Data
Control of Behaviour: with 77 figures/Frederick Toates (ed.)
Includes bibliographical references and index.
ISBN 3-540-63795-8 (pbk.)
1. Animal behaviour. I. Toates, F.M. (Frederick M.)
QL751.C675 1998
591.5—dc21 97-52174 CIP

Published by Springer-Verlag, written and produced by The Open University

Cover design: *design & production* GmbH, Heidelberg

Printed in Singapore by Kyodo under the supervision of MRM Graphics Ltd, UK.

ISBN 3-540-63795-8 Springer-Verlag Berlin Heidelberg New York

This text forms part of the Open University *Biology: Brain & Behaviour* series. The complete list of texts which make up this series can be found above. Details of Open University courses can be obtained from the Course Reservations and Sales Office, PO Box 724, The Open University, Milton Keynes MK7 6ZS, United Kingdom: tel. (00 44) 1908 653231. Alternatively, much useful course information can be obtained from the Open University's website: http://www.open.ac.uk

2.1

SPIN 10654267 #39/3137 – 5 4 3 2 1 0

CONTENTS

PREFACE

Control of Behaviour, like any other textbook, is designed to be read on its own, but it is also the fifth in a series of six books that form part of *SD206 Biology: Brain and Behaviour,* a course for Open University students.

Each subject is introduced in a way that makes it readily accessible to readers without any previous knowledge of that area. Questions within the text, marked with a □, are designed to help readers understand and remember the topic under discussion. (Answers to in-text questions are marked with a ■.) The major learning objectives are listed at the end of each chapter, followed by questions (with answers given at the end of the book) which allow readers to assess how well they have achieved these objectives. Key terms are identified in bold type in the text; these are listed, with their definitions, in a glossary at the end of the book. Key references are given at the end of each chapter, where appropriate. A 'general further reading' list, of textbooks relevant to the whole book, is also included at the end.

The study of the brain and behaviour is an experimental science. This means that it involves the collection of observations, the formulation of specific hypotheses to explain those observations and the carrying out of experiments to test (confirm or falsify) those hypotheses. Throughout this book, these different aspects of the investigative process are emphasized, often through the use of in-text questions in which the reader is invited to engage in the process of deductive reasoning themselves. An understanding of the scientific method, as it applies to the behavioural and brain sciences, is an important aim of this book.

This book describes those factors internal and external to the animal that control behaviour. For example, it will consider the factors underlying the change from being active to being asleep, the switch from one activity to another and the change in responsiveness to a constant stimulus. The first chapter outlines the various approaches to the study of the control of behaviour. Chapter 2 describes the study of motivation as a way of explaining the control of behaviour. The rhythmic cycles of behaviour, such as the sleep cycle, form the core of Chapter 3. In Chapter 4, one particular behaviour, aggression, is considered for the point of view of a causal and a functional explanation. Aggression was taken as an example, not because it is especially relevant to a study of the control of behaviour, but because it has attracted a lot of attention, notably from behavioural and social scientists.

When behavioural systems are stretched beyond their range of optimal functioning, the result is stress. Chapter 5 examines stress in both causal and functional terms.

Before you begin to read this book, there are some important points that you should bear in mind.

1 Experiments on animals

The use of living animals in research is a highly emotive, contentious and political issue. You are no doubt aware of the strong views held by animal liberationists. There is also considerable debate among scientists concerning what kinds of experiments and procedures are acceptable and what are not. Most scientists working with animals seek to minimize any suffering that animals may experience during experiments and each researcher makes his or her own judgement as to whether the suffering caused by an experiment is justified by the scientific value of the results that the experiment yields. The ethics of animal experimentation is not simply a matter of individual judgement, however, but is a matter of concern for society as a whole. In Britain and many other countries, all researchers work within strict guidelines enforced by government; for example, the Home Office licenses all animal experimentation in the UK. Some academic societies, such as the Association for the Study of Animal Behaviour, and many institutions, such as medical schools, have Ethical Committees that oversee animal-based research. In this book, a number of experiments are described; this in itself raises ethical issues because reporting the results of an experiment may be thought to be giving tacit approval to that experiment. This is not necessarily true and it should be pointed out that some of the experiments described were carried out several years ago and a number of them would not be carried out today, such has been the shift in opinion on these issues within the biological community. Paradoxically, certain experiments carried out many years ago, such as those on the effects of maternal deprivation on young monkeys, produced such strong and distressing effects on their subjects—results that were not generally anticipated—that they have had a substantial impact on the kind of experiments that are permitted today.

2 Latin names for species

A particular individual animal belongs to various categories. If you own a pet, it may, for example, be categorized as a bitch, a spaniel, a dog, a mammal, or an animal. Each category is defined by particular features that differentiate it from other, comparable categories. The most important level of categorization in biology is at the level of the species. When a particular species of animal is referred to in this book, its Latin name is also given, e.g. earthworm (*Lumbricus terrestris*).

CHAPTER 1
INTRODUCTION

This book describes those factors, internal and external to the animal, that control behaviour. Consider what needs to be explained. An animal might be mating one moment and then switches to drinking. What determines this switch in behaviour? Over a longer time period, an animal might spend 12 hours sleeping and then 12 hours in activity of various kinds. This book will ask what are some of the factors that underlie such a change in activity.

Consider again the example (Book 1, Chapter 7) of a rat housed in a laboratory cage with food and water available *ad libitum* (freely). Even if such an environment is held constant, behaviour will not be constant. The rat switches between sleeping, grooming, feeding and exploring. This book is concerned with explaining the processes that underlie such switches.

Animals sometimes respond to identifiable stimuli in the environment, e.g. a mate or a potential item of prey. In some cases, behaviour is a fairly predictable consequence of certain stimuli, e.g. a sudden loud noise will cause birds feeding on the ground to take to the air. At other times a given stimulus might or might not evoke a particular response. For example, a female rat whose hypothalamus has been sensitized by oestrogen will respond to tactile stimulation from the male with a lordosis response (Book 2, Chapter 10). At other times she will not respond. This book is concerned with explaining the variability of the response to external stimuli.

Researchers into the control of behaviour have various affiliations, e.g. psychology, ethology and neurophysiology. Therefore, this book, like Books 3, 4 and 6, has an interdisciplinary approach to the topic. Within the disciplines there are different starting points for gaining understanding. Chapter 2 will describe the study of motivation (Book 1, Chapter 7) as a way of explaining the control of behaviour. It considers some of the neurobiological bases of motivation, and calls upon knowledge of neurons and neurotransmitters that was introduced in Book 2.

Chapter 3 looks at rhythms. You have already met various examples of rhythms in the course, e.g. the oestrous cycle underlying mating. Such cyclic variations in hormones constitute a varying internal factor that has a powerful control over behaviour. A number of rhythms and their properties are examined, e.g. the rhythm of sleeping and waking.

Chapter 4 looks at one particular behaviour, aggression, and considers both causal and functional levels of explanation. In principle, any behaviour could have been chosen to be described closely in this way. One could look at drinking, in terms of the physiological benefits of, for example, fluid gain at a time of dehydration and the costs of spending time engaged in drinking at an exposed water-hole. However, aggressive behaviour and sexual behaviour have perhaps attracted most attention from ethologists in terms of causal and functional explanations.

Chapter 5 builds on the earlier chapters in Book 5 to consider what happens when behavioural systems are stretched beyond their range of optimal functioning. The chapter will examine the external and internal controls under these conditions.

Book 5 has a close association with Book 3, which looked at communication. To some extent, Book 5 is also about communication; it is concerned with such things as how information on nutrients leaving the gut is communicated to the brain in the control of feeding, how information on the size of a rival influences the decision to fight, and how information at the level of the hypothalamus is communicated by hormones to the adrenal gland. However, the emphasis in this book is on *action*. How is information concerning events external and internal to the animal translated into behaviour?

Finally, Book 5 leads logically into Book 6, which is concerned with the effects of degeneration, damage and disorder on brain and behaviour. The final chapter of Book 5 deals with stress, and so forms a natural interface between, on the one hand, the control of behaviour under 'normal' conditions and, on the other, pathology.

Different *levels of explanation* will be offered in this book. You have already seen something of these various levels of gaining understanding. Some researchers work at a purely behavioural level, devising models and accounts of behaviour that do not address the underlying neural processes. On the other hand, some researchers are interested in the activity of neurons and neurotransmitters, with little or no regard to behavioural events.

The broad consensus of opinion, as reflected in the present course, is that scientisis need to look at various levels to gain insight into brain and behaviour. They also need to attempt to relate the levels to one another. So, for example, it is useful if those processes that are postulated by psychologists and ethologists to underlie behaviour can be given some kind of embodiment in real nervous system structures and processes. This requires researchers sufficiently 'fluent' in both psychology/ethology and neural science so as to be able to speak both 'languages'. In the present book, Chapter 3 will look at two ways of approaching biological rhythms: a so-called black-box approach (Book 1, Section 2.2) in which the rhythms are investigated without reference to their neural basis, and the approach of looking for the basis of the rhythm in the nervous system.

From Book 1, Chapter 7 you will already be familiar with motivation. Rhythms will be less familiar to you and you will not have met the relationship between rhythms and motivation in the control of behaviour. The following section introduces this topic so that you will be able to relate the material on motivation to that on rhythms and place them into the broader context.

1.1 Biological rhythms, motivation and the control of behaviour

Over a period of time, it is common for some behaviour patterns to vary rhythmically. For example, female rats are only sexually receptive for a period of 12–18 hours every 4–5 days. Depending upon the species, mating, feeding and drinking are most likely to occur in either the light or dark phase of the 24-hour period. Typically, animals show 24-hour rhythms in sleeping/waking. Such rhythms are ubiquitous in biological systems.

1.1.1 Feeding and drinking

Feeding and drinking perhaps provide the clearest examples of where, in order to understand the control of behaviour, it is necessary to come to terms with both motivation and rhythms. Figure 1.1 shows the occurrence of meals and drinks for a laboratory rat during each 24 hours over a period of 7 days. It is clear that most meals and drinks occur in the dark phase and relatively few in the light phase. Similarly, if general activity is measured, rats are more active in the dark than in the light.

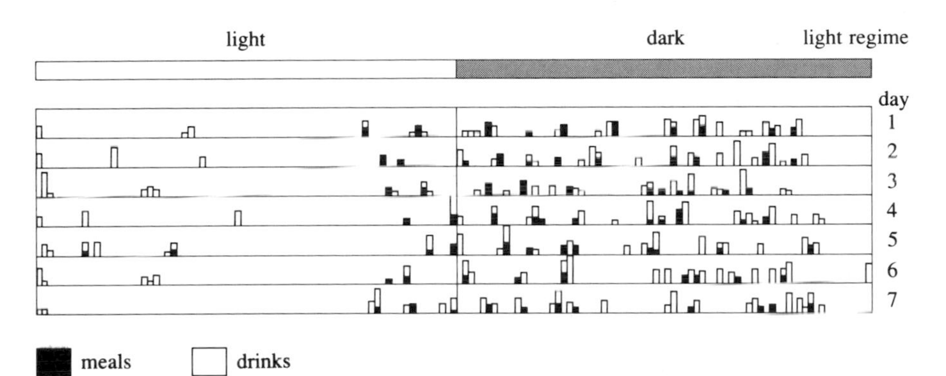

Figure 1.1 Feeding and drinking pattern over a 7-day period of a rat housed in a cage exposed to a light regime of 12 hours dark/12 hours light. Food and water were freely available throughout. The height of each bar indicates the amount ingested.

- ☐ Considering energy balance, give a possible reason for this difference in feeding between light and dark phases.
- ■ Increased activity in the dark would require more energy and this might explain why more food is ingested in the dark.

Alternatively, there might be no variation in metabolic need over the 24 hours, but there might be changes occurring within the feeding system that underlie the rhythm. If the latter is true, the animal might be storing energy in some form in its body during the dark and it might utilize this in the light (the nature of the food and the construction of the cage precluded hoarding of food). Metabolic rate is a measure of the rate at which the body is consuming energy. Figure 1.2 shows the variation in both energy intake and metabolic rate for a group of rats over a 24-hour period. The graph represents the average (mean) of 30 day/night cycles, with the result that each individual rat's typical nocturnal eating pattern of several distinct meals with pauses between has been ironed out. Rats, of course, do not eat at a steady rate throughout the night.

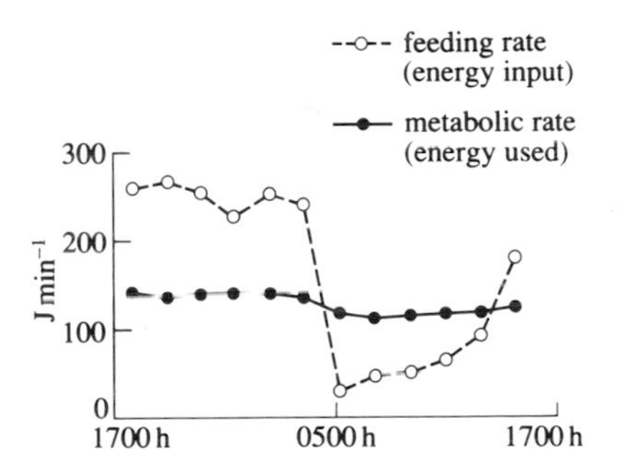

Figure 1.2 The amount eaten by rats and their metabolic rates (values were the mean of 30 day/night cycles). Values are given in terms of energy intake and energy used, respectively, and were measured in Joules min^{-1}. Lights were off between 1700 h and 0500 h, and on between 0500 h and 1700 h.

- ☐ Can variations in metabolic rate account for the variation in food intake?
- ■ No. Although there is a slight variation in metabolic rate, this is insufficient to explain all of the variation in feeding.

The result strongly suggests that some of the food eaten during the dark phase is utilized for metabolism during the light phase when the animal eats very little. To test this, researchers radioactively labelled some food in order to 'tag' it. It was found that food ingested during the light phase was metabolized within the same phase, whereas a large part of the food ingested during the dark phase was converted in the body into fat, stored and used as energy during the light phase. Thus, variations in metabolic rate cannot account for variations in food intake.

In some species, rather than storing energy inside the body, it can be stored either in mouth pouches or, in an appropriate environment, outside the body entirely, i.e. in a hoard. The golden hamster (*Mesocricetus auratus*) can store food both ways. Table 1.1 shows the light/dark rhythm of food-related activity in the hamster. The amount taken from the food dispenser was carefully measured, as was the size of the hoard in the light and dark phases.

Table 1.1 Feeding and hoarding behaviour of hamsters

	Amount taken from source/g	Change in hoard size/g	Amount dropped on floor/g	Calculated amount eaten/g
light	1.2	−2.5	0	3.7
dark	12.1	+5.8	1.5	4.8

□ What does Table 1.1 show?

■ That during the dark the hamster is making a net increase in the size of the hoard, whereas during the light it is making a net withdrawal.

Thus, to understand fully the control of feeding, it is necessary to know about factors such as homeostasis and how the energy state influences feeding. In addition, it is necessary to know about the important role that rhythms play.

Consider now the light/dark rhythm of drinking, also shown in Figure 1.1.

□ There are at least two possible explanations that might be given for this rhythm. Cast your mind back to Book 1, Chapter 7 and see whether you can come up with them.

■ (1) There could be an intrinsic rhythm in the drinking system, or (2) there might be no intrinsic rhythm in the drinking system, with increased drinking in the dark phase being explainable in terms of the increased feeding taking place then, since food intake is known to exert an excitatory effect upon drinking (Book 1, Section 7.2.2).

□ What is a possible experiment that could reveal whether there is an intrinsic rhythm in the drinking system?

■ The rhythm in feeding could be ironed out by first depriving the animal of food for a while and then allowing it to take meals only at regular spaced intervals.

The result of such an experiment, together with that for the period of *ad libitum* feeding and drinking that immediately preceded it, is shown in Figure 1.3.

□ Which of the two hypotheses given above is supported?

■ Even when meals are taken at regular spaced intervals, most drinking still occurs in the dark phase. This suggests that the drinking rhythm is not just a consequence of the rhythm in feeding.

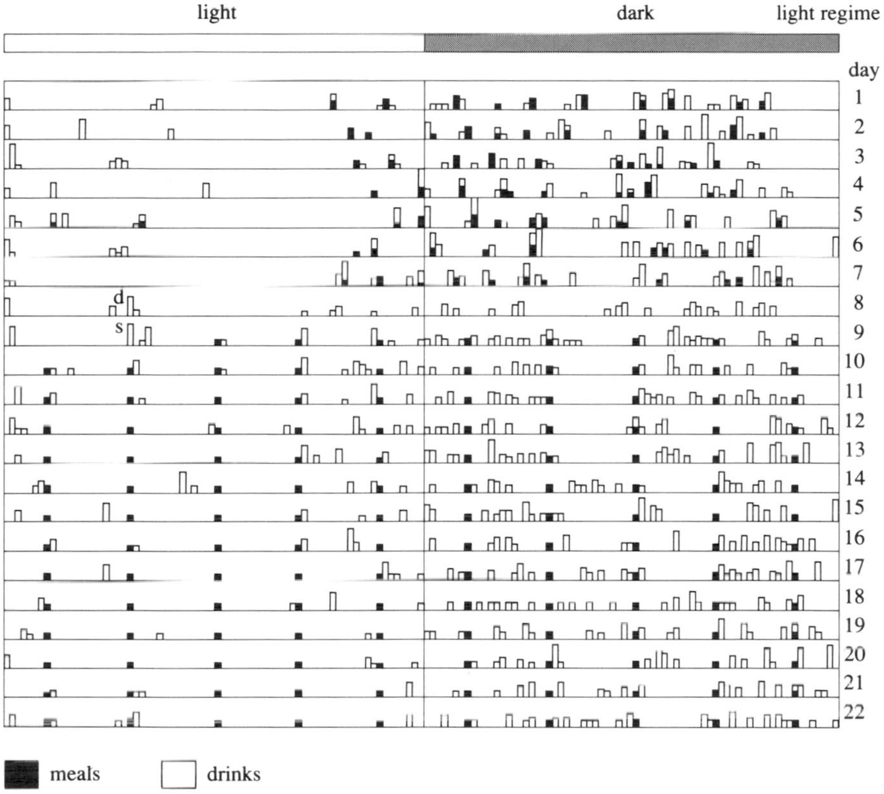

Figure 1.3 Feeding and drinking pattern of a rat housed in a cage exposed to a light regime of 12 hours dark/12 hours light. Water was available *ad libitum* throughout. For the first 7 days, food was also available *ad libitum*. The height of each bar indicates the amount ingested. For the period starting at 'd' and ending at 's', the rat was deprived of food. Starting at 's', meals were made available (and were consumed by the rat) only at fixed and regular intervals throughout the 24-hour period.

The existence of light/dark rhythms in feeding and drinking is well established. How are they to be explained? The rhythm could be the direct result of the variation in light intensity. Light falling on the eyes might inhibit feeding and drinking.

□ How could you test this?

■ Place rats in either constant dark or constant light and see whether a rhythm still persists.

The results of such experiments are discussed in Chapter 3. For the moment, suffice it to say that a clear rhythm persists, even under these conditions.

Therefore, the rhythm is *intrinsic* to the animal, rather than being directly dependent upon stimuli from the environment.

1.1.2 Homeostasis, negative feedback and rhythms

The fact that, if a disturbance is imposed upon a physiological negative feedback system, the system takes corrective action to maintain the variable at the undisturbed value, does not mean that such variables are necessarily held at a constant value over time. For example, as Chapter 3 will show, human body temperature exhibits a cycle during a 24-hour period. The temperature regulation system itself adjusts body temperature in this cyclic fashion.

Section 7.2.2 in Book 1 introduced an analogy between the body temperature regulation system and a home heating system which regulates room temperature at a set-point value. To extend this analogy, room temperature might vary for several reasons. Two possibilities are: (1) variation in the weather for which the control system would make some compensatory action, or (2) variation because the occupant of the room raises and lowers the set-point. The latter is analogous to the cycle of body temperature in that the source of variation arises from inside, rather than outside, the system.

There is a fundamental difference then between (1) change imposed from *outside* which is automatically resisted, and (2) change instigated *internally within the control system*, which is not resisted. Thus, if the weather starts to move body temperature down, corrective action is taken. If the control system itself does so, for example as part of the 24-hour rhythm, compensatory action is not elicited.

1.1.3 Functional aspects of negative feedback and rhythms

Rhythms seen in feeding, drinking and other behaviour patterns are intrinsic to the animal. Expressed in other words, the rhythm serves as an internal model of the day/night cycle rather than being dependent upon it. In effect, it allows the animal to tell the time. Thus, for example, a rat burrowed beneath the earth can emerge at dusk and does not have to perceive directly the fall in illumination to do so. By analogy with the animal's intrinsic rhythm, a wrist-watch serves as a model or representation of the external world. In contrast to a wrist-watch, a sundial is directly dependent upon external illumination and is useless at night or on a cloudy day. Therefore, Chapter 3 'Biological clocks' is appropriately titled. Of course, no one is suggesting that animals consciously go to consult a clock as to what to do. Rather, some kind of cyclic influence acts upon their behaviour.

Viewed in functional terms, the negative feedback process and rhythms underlying behaviour reflect two different adaptations for coping with two different aspects of the environment—its unpredictability and its predictability, respectively. Aspects of an animal's environment can change day-to-day from being relatively constant to being highly variable. Some changes within an environment will be unpredictable (e.g. a sudden flood, a fall of snow in June), whereas others will be predictable (e.g. the repeated cyclic nature of changes in light and dark). In negative feedback, action is taken in response to *deviations from normal* and aids survival in the face of either predictable or unpredictable disturbances. For example, a sudden snowfall in July can result in processes of temperature conservation being recruited. Behavioural temperature control is utilized to keep warm. In other words, action is taken *in response to* a deviation; but, of course, in a negative feedback system, the deviation has to be suffered first before action can be taken to correct it.

As efficient as such negative feedback systems are at correcting disturbances, it is not difficult to appreciate that an animal that has processes that can, in effect, *anticipate* what is happening in the internal and external environment, and that can cause appropriate action to be taken, will be at an advantage over one that simply responds to events as they happen. If the deviation can be anticipated, action can be taken to pre-empt it.

□ You have met an example of anticipation already in the course. What was it?

■ Section 7.2.2 in Book 1 described the feedforward system by which rats anticipate the dehydration that can follow eating a meal. By drinking in association with the meal, rats pre-empt such dehydration.

In effect, rhythms provide another means of anticipation. In evolutionary terms, they could only have arisen in an environment in which the feature with which they are associated, such as a 24-hour light/dark cycle, is highly predictable. In the natural environment, the rhythm underlying the behaviour of a nocturnal animal that ventures out from a burrow to feed at dusk is anticipating the future need for energy. The animal stocks up with nutrients at a time when it is most adaptive to forage.

□ Give two examples of what such stocking up can consist of.

■ In rats, food is eaten in excess of metabolic need during the dark phase, and is utilized for metabolism during the light phase. In hamsters, food is added to the hoard during the dark phase and taken from the hoard and eaten in the light phase.

In both cases, nutrients gathered in the dark phase will last the animal through the light phase. Hence, the animal feeding or hoarding at night is not responding simply to the current deficit in energy since it is acquiring nutrients in excess of metabolic needs. Rather, the rhythm is exerting a control over its behaviour in a way that anticipates future need.

Therefore, both negative feedback and anticipatory aspects serve an adaptive role. Both maintain the value of important physiological variables. In doing so, they play complementary roles, each reflecting different aspects of the environment. The control exerted over ingestion by the intrinsic rhythm allows the animal to cope with the predictable feature of the environment, alternation between dark and light. The control exerted by negative feedback allows the animal to cope with the unexpected.

Perhaps the best known rhythm is that of sleep and waking, discussed in Chapter 3. Sleep is a phenomenon found in many species and yet its functional value is uncertain. What is certain is that humans usually feel awful after a period of deprivation from sleep, but quite what is happening during sleep remains unclear. One view is that sleep evolved in the service of a homeostatic function; in this view, during the hours of waking a substance or substances builds up or is depleted in the body. Sleep serves to restore the level of this substance to normal. Others argue that sleep evolved as a mechanism to keep an animal out of action when it is most vulnerable. Primates tend to be vulnerable in the dark, when the chances of being attacked by a predator are greatest. Therefore, a powerful motivation to find a safe place to hide in during the dark phase has evolved. Other animals (e.g. rats) are most vulnerable in the light.

These need not be mutually exclusive functional explanations. Sleep could have evolved both to keep the animal out of danger and to allow, whilst asleep, some kind of restoration of a biochemical state, e.g. a biochemical change in neurotransmitter levels in the nervous system. Even if one accepts the hypothesis of keeping the animal out of danger, there still needs to be a causal explanation in terms of, one imagines, some biochemical change that alternately plays a role in sending it to sleep and waking it up.

Summary of Chapter 1

Rhythms exert control over behaviour by acting upon the control system itself. For example, the cyclical changes in human body temperature over a 24-hour period are a property of the temperature control system. It should be distinguished from changes in body temperature arising from outside the system, for example, from changes in the weather. Such externally imposed changes evoke corrective action that tends to cancel their effects. By contrast, changes intrinsic to the system are not opposed by the system.

In negative feedback systems, action is taken in response to a disturbance. Such systems are adaptations that serve well to counter unexpected disturbances. Rhythms are adaptations that reflect predictable cycles such as the light/dark cycle over each 24-hour period. They allow the animal to anticipate. In rodents, such as rats, feeding and drinking occur mainly in the dark period. Although metabolic rate is slightly higher in the dark period, this cannot account for the magnitude of the change in feeding. Rather, food gathered (and either eaten or hoarded) during the dark period is utilized in the light period. By stimulating eating when this activity can best be carried out, sufficient food can be ingested or hoarded to cover metabolic needs during the light period.

Objectives for Chapter 1

When you have completed this chapter you should be able to:

1.1 Describe some of the properties of rhythms in feeding and drinking. (*Questions 1.1 and 1.2*)

1.2 Describe the functional significance of the control exerted by negative feedback and by rhythms. (*Question 1.3*)

Questions for Chapter 1

Question 1.1 (*Objective 1.1*)
In the experiment shown in Figure 1.3, why did the experimenter impose a period of food deprivation between the period when food was available *ad libitum* and the period when the animal was placed on a schedule of spaced meals?

Question 1.2 (*Objective 1.1*)
The evidence strongly suggests that there is an intrinsic rhythm within the feeding control system. However, to be certain, how would you test whether the rhythm in feeding is intrinsic to the feeding motivation system or driven by the rhythm in drinking?

Question 1.3 (*Objective 1.2*)
Two hypothetical students are having a discussion, part of which is reproduced below. Describe where Jill and Ali are going wrong.

Jill I thought behaviour was all about defending against disturbances and keeping the conditions in the body constant. I can't see how this can be the case when these same systems let rats eat and drink everything at night and nothing by day.

Ali The body has various negative feedback loops and they are good at coping with small disturbances, like when you go into a cold room and start to shiver, but they can do nothing to stop some really big disturbances like those seen in rhythms.

Further reading

Armstrong, S. (1980) A chronometric approach to the study of feeding behaviour, *Neuroscience and Biobehavioural Reviews*, **4**, pp. 27–53.

CHAPTER 2
MOTIVATION

2.1 Introduction

Traditionally, the study of the factors that determine an animal's various activities has fallen largely within the area of psychology and ethology termed 'motivation' (Book 1, Chapter 7). The essence of the study of motivation is the question of how internal and external factors contribute to motivational processes within the animal, and thereby to behaviour. Therefore, the central theme of this chapter will be how the study of motivation can contribute to an understanding of the control of behaviour.

Figure 2.1 shows just two of the many motivational systems that underlie behaviour, those of feeding and drinking. Each motivation depends upon both external and internal factors. Expression of a motivation in behaviour is associated with goal-directed (often termed 'appetitive') behaviour towards either food or water, and with consummatory behaviour.

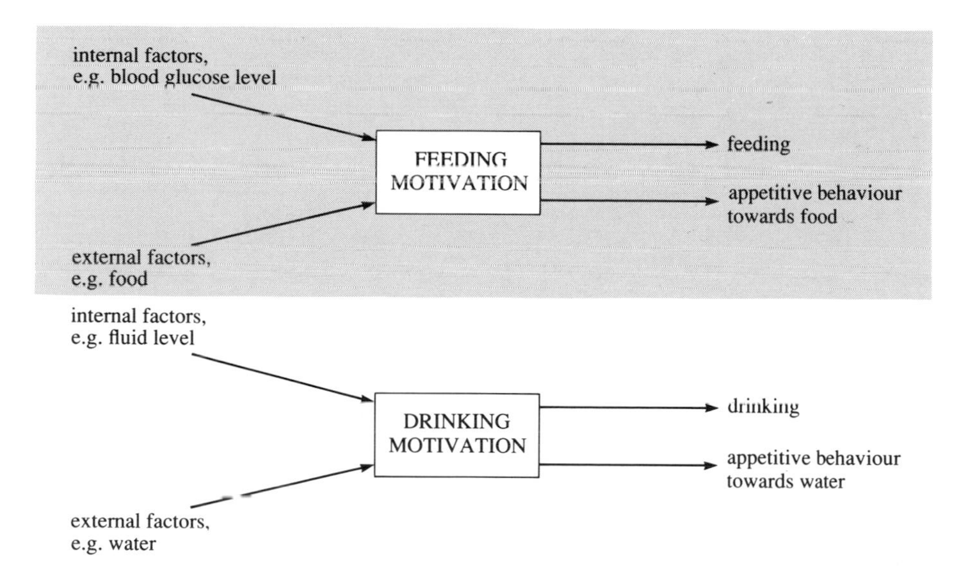

Figure 2.1 Two motivational systems. Each motivation depends upon external and internal factors. When a particular motivational state, e.g. feeding motivation, is expressed in behaviour, an animal shows either appetitive behaviour towards food, or feeding.

Section 2.2 will look at a single motivational system, that underlying feeding, and will show how internal and external factors jointly determine behaviour. In terms of Figure 2.1, the section draws a boundary around the feeding system and looks at it in isolation. The experiments described were carried out in an environment designed to minimize 'distractions' from other motivations. It was shown in Book 1, Chapter 7 that feeding motivation depends, amongst other things, upon (a) the internal state of the animal (e.g. level of blood glucose), and (b) the available food. Some researchers have probed the nervous system to look for neurons whose activity reflects these aspects of motivation, i.e. the neurons' activity would be sensitive to both internal state and external food-related stimuli. Deprivation of food might be expected to increase the firing of such neurons. Section 2.2 will look at some of the neural processes underlying feeding.

Book 1, Chapter 7 and Book 2, Chapter 10 discussed not only the behavioural aspects of, for example, feeding and mating, but also the goal-directed (or 'appetitive') behaviour that leads to contact with such stimuli as food or a mate. Section 2.2 will pursue this theme in the case of feeding. It will show how procedures can affect motivation, as measured by food intake (consummatory behaviour), but not by the behaviour of getting to food (appetitive behaviour).

Building on the understanding of individual motivational systems, Section 2.3 will look at some of the processes involved in behavioural decision making. Figure 2.2 develops Figure 2.1 by taking into account some of the interactions between systems. For example, in the case of the incubating hen (Book 1, Section 7.6), there is an inhibition from the system of incubation to that of feeding, shown as I_1. Just as an incubating hen cuts down on her food intake, so do a wide variety of species when water is not available. This happens even in the midst of food. Thus Figure 2.2 shows an inhibitory connection (I_2) from the drinking motivational system to that of feeding.

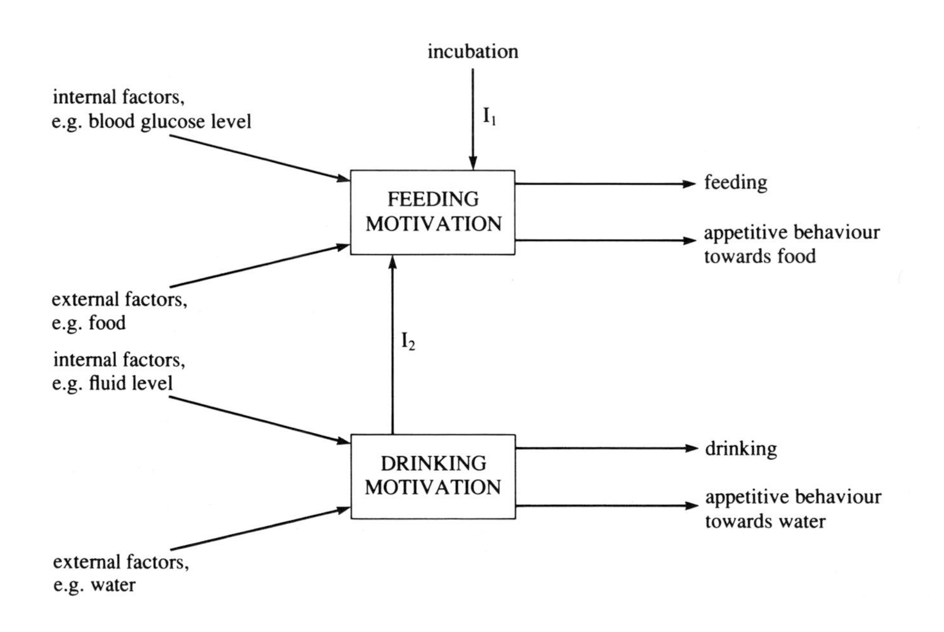

Figure 2.2 Development of Figure 2.1, showing two possible interactions between feeding and other motivational systems. Incubation and drinking motivation exert inhibitions (I_1 and I_2, respectively) on feeding motivation.

There is another more obvious way in which the study of behaviour in anything but a simple environment requires looking at interactions. The situation is often such that animals can only do one thing at a time. To take an obvious example, turning left towards a place where there is food would be incompatible with turning right towards a place where there is water. Thus, if appetitive behaviour towards water is being performed, inhibition is exerted upon the tendency to show appetitive behaviour towards food. Section 2.3 will look at some examples of such inhibition between systems. Thus, necessary as it is sometimes to simplify and study systems in isolation, in other contexts it is necessary to acknowledge that motivational systems do not exist in isolation, and to look at the interactions between them.

Section 2.4 also considers feeding, this time in the context of the role of neurotransmitters implicated in the processes of motivation. Sometimes researchers attempt to manipulate motivation by artificial means that target neurotransmitters. For example, there is commercial pressure to develop a drug that would lower appetite for food. Physiological psychologists would need to ask a number of questions of such a drug. For example: Where in the body does it act, on the brain or elsewhere? What are its undesirable effects (usually termed 'side-effects')?

Section 2.5 looks at drug-taking, specifically the opiates and cocaine. It shows how insight can be gained by looking at such behaviour in terms of a broader perspective of motivation and by applying the kinds of theoretical model that are developed earlier in the chapter for understanding feeding and other systems.

2.2 Internal and external factors in the control of feeding

Section 2.2 looks at the feeding system in isolation from other systems. Figure 2.3 shows, as a possible first simplification, the main sets of factors that determine feeding. First, there are internal metabolic factors, such as the level of glucose in the blood. There is also a daily rhythmic influence on feeding (Section 1.1.1). Second, there is the primary external factor, food. Foods vary in their palatability; some are ignored while others are readily ingested. Third, there are external stimuli that have been paired in the past with food presentation, i.e. conditional stimuli (Book 1, Section 7.4.1).

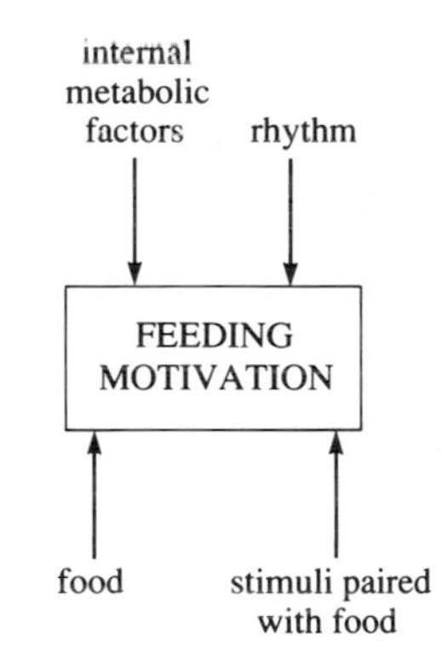

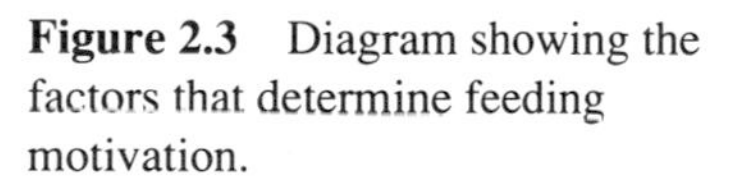
Figure 2.3 Diagram showing the factors that determine feeding motivation.

The power of food to arouse feeding cannot be understood simply in terms of the physical characteristics of the food. Rather, it is sometimes necessary to look to events within the animal's history to understand its reaction to a particular food.

□ Can you give examples of where this effect is seen?

■ Two possibilities are:
1 The power of a particular flavour to arouse ingestion depends to some extent upon its novelty value (Book 1, Section 7.3). Novelty in this context is defined in terms of what the animal has eaten recently.
2 The capacity of food to arouse feeding is decreased if, in the past, it has been associated with nausea (Book 1, Section 6.4.1).

The remainder of this section will look more closely at some of these features of motivation, in the case of feeding. First, the detection of the internal metabolic state of the animal is discussed.

2.2.1 Detection of the metabolic state

Depriving an animal of food for a period of time is a powerful way of increasing feeding motivation. Food deprivation is associated with a depletion of the animal's reserves of energy-providing substances, and the blood glucose level may fall. The task of locating the detectors of the body's energy state that play a role in the control of feeding has presented formidable difficulties, with first one site after another being 'discovered' as *the* detector.

It now seems that signals arise from events at more than one site, and the internal signal that contributes to feeding arises from an integration based on the signals from these various sites. The hypothalamus (Book 2, Section 10.3) plays a crucial role in this integration. At the moment, the evidence points to the brain and the liver as being the two most important sites for the detectors. Glucoceptive neurons (i.e. neurons that are specifically sensitive to the presence of glucose) have been located in particular regions of the hypothalamus. Their rate of generating action potentials increases as the concentration of glucose in their vicinity increases. Another class of neuron has been located in another region of the hypothalamus; this class decreases its rate of action potentials as glucose concentration increases.

The liver would seem to be an appropriate site for a detector of energy state (see Figure 2.4). Nutrients from ingested food are absorbed into small blood vessels—the capillaries—that line the walls of the intestine. They then travel, via the hepatic portal vein, straight to the liver. The location of the liver is therefore ideal for monitoring the supply of nutrients from the gut. According to contemporary theories, when this nutrient supply falls, feeding tends to be triggered. In rats, infusions of glucose into the hepatic portal vein are particularly effective in inhibiting feeding, compared with infusions made elsewhere, but the power of such infusions to switch feeding off is lost if the neurons carrying information from the liver to the brain are cut. Reductions in the frequency of action potential generation of neurons carrying information from the liver to the brain have been recorded in response to increases in concentration of glucose in the hepatic portal vein; the reductions were in proportion to the increases in concentration.

It was mentioned earlier that there are neurons in the hypothalamus whose rate of firing is reduced by increases in glucose concentration in the fluid that bathes them. A reduction in their firing rate has also been noted in response to glucose infusions into the hepatic portal vein, suggesting that such neurons are performing an integrative function. They are activated by a low level of glucose either in their immediate vicinity or at the liver, or both.

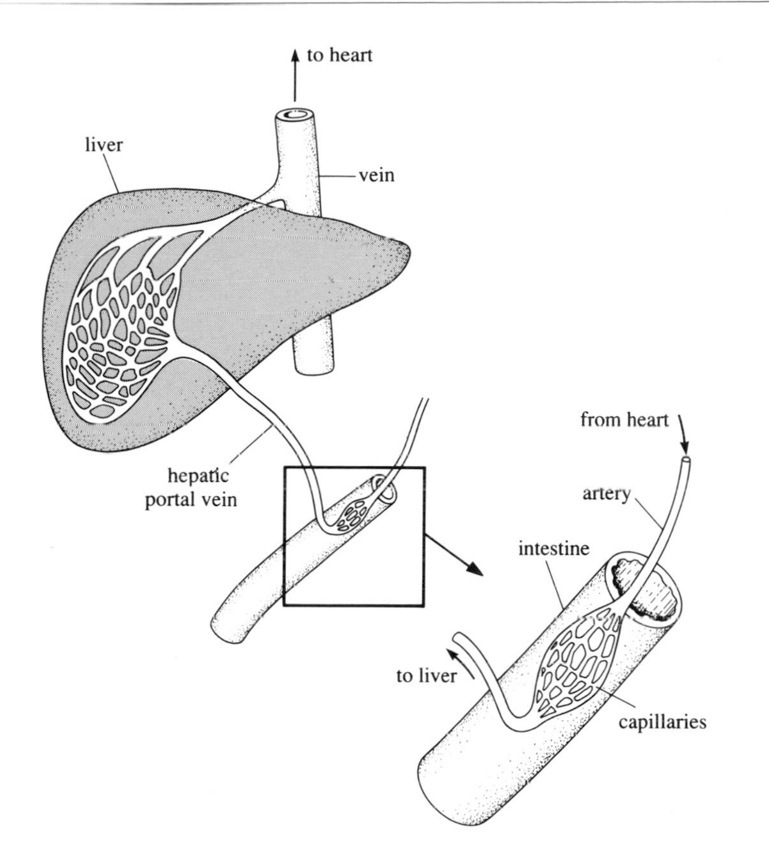

Figure 2.4 Nutrients are absorbed by capillaries that line the wall of the intestine, and are then carried in the hepatic portal vein to the liver.

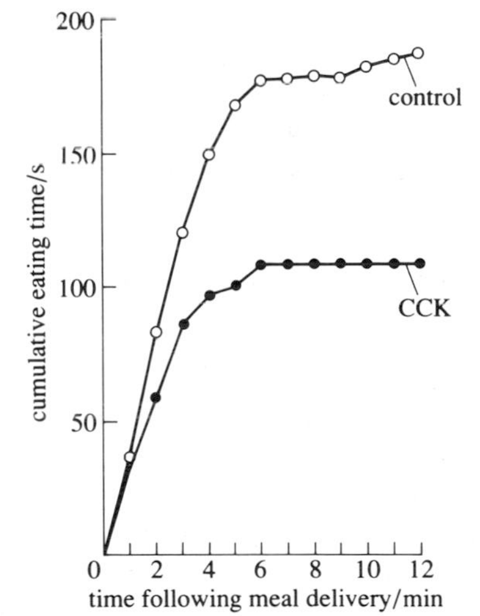

Figure 2.5 Cumulative time spent eating following food being made available, by rats injected either with CCK ($11.2\,\mu g\,kg^{-1}$) or saline. Injected saline was at the same concentration as the body fluids, to act as a control for the injection itself.

2.2.2 Short-term feedback

Feeding is similar to drinking in that one of the major instigating events is a change in a physiological variable within the body (Book 1, Section 7.2.2). However, following the onset of feeding, it takes time for any such change to be reversed. In an animal such as the laboratory-housed rat, the act of drinking or feeding is normally completed before the physiological change would have time to occur. This implies that there are short-term negative feedback loops that serve to switch off eating or drinking before the normal physiological state has been restored.

This *short-term feedback* arises from the mouth and stomach to terminate drinking and feeding. In the case of feeding, the hormone cholecystokinin (CCK) is implicated in this capacity. Food in the gut causes the release of CCK, which triggers a sequence of neural activity that exerts an effect in the brain. Injections of CCK have the effect of reducing feeding. An example of such an effect of CCK is shown in Figure 2.5.

□ CCK reliably lowers the amount of food eaten. What caution is necessary in using this result to make conclusions about the normal function of CCK?

■ Injected CCK might reliably lower feeding but might do so by some means unconnected with what naturally terminates feeding, e.g. it might make the animal feel sick, or sleepy.

To answer this, researchers have tried to rule out, for example, a general sedation effect of CCK. If the action of CCK is specific to feeding, its injection should not, for example, lower drinking in animals following water deprivation. In rats deprived of water, CCK lowers the intake of a nutrient-containing liquid diet but not the intake of pure water. Thus, its satiety role seems specifically to involve the detection of nutrient intake, and does not involve general sedation.

2.2.3 Appetitive and consummatory aspects of feeding

The conditioning technique of H. P. Weingarten (1984; see Book 1, Section 7.4.1), has been utilized to examine feeding more closely.

□ What was Weingarten's experiment?

■ A neutral stimulus (e.g. a tone) was paired with food presentation when the rats were hungry. Later, when the rats had been returned to *ad libitum* feeding conditions, the tone was shown to have the capacity to elicit feeding.

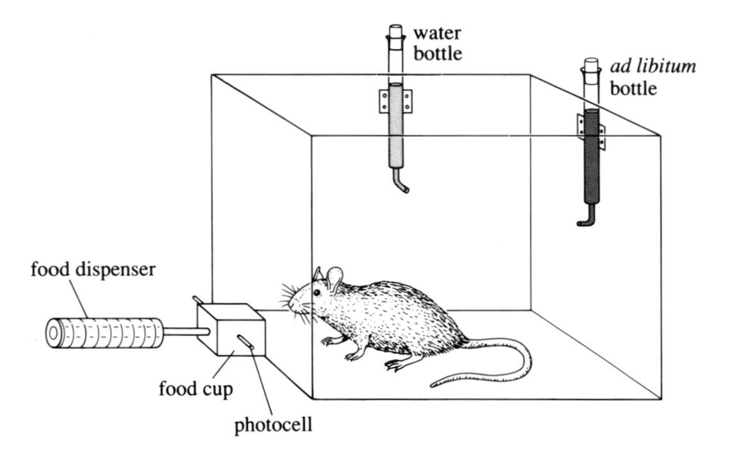

Figure 2.6 Apparatus employed by Weingarten. During the training period, food was only available from the cup. During testing, liquid food was available from the *ad libitum* bottle and, under the control of the experimenter, an identical liquid food was available from the cup. Nose-pokes (investigations of the food cup by the rat) into the food cup were measured by a photoelectric cell.

The apparatus is shown in Figure 2.6 and the procedure is summarized in Figure 2.7. First, there was the *ad libitum* period (1) during which the rats were fed *ad libitum* and adapted to the apparatus. This was followed by the training period (2), during which rats were first deprived of food for a period. Then, a tone and light (termed tone_1 + light_1) were presented for 4.5 minutes and a meal delivered in a food cup during the last 30 s of the 4.5 minutes (shaded area). The combination of tone and light was termed a conditional stimulus (CS), since any later capacity to elicit feeding in satiated rats depended upon its earlier pairing with food. It was designated as CS+, meaning the stimulus that signals food.

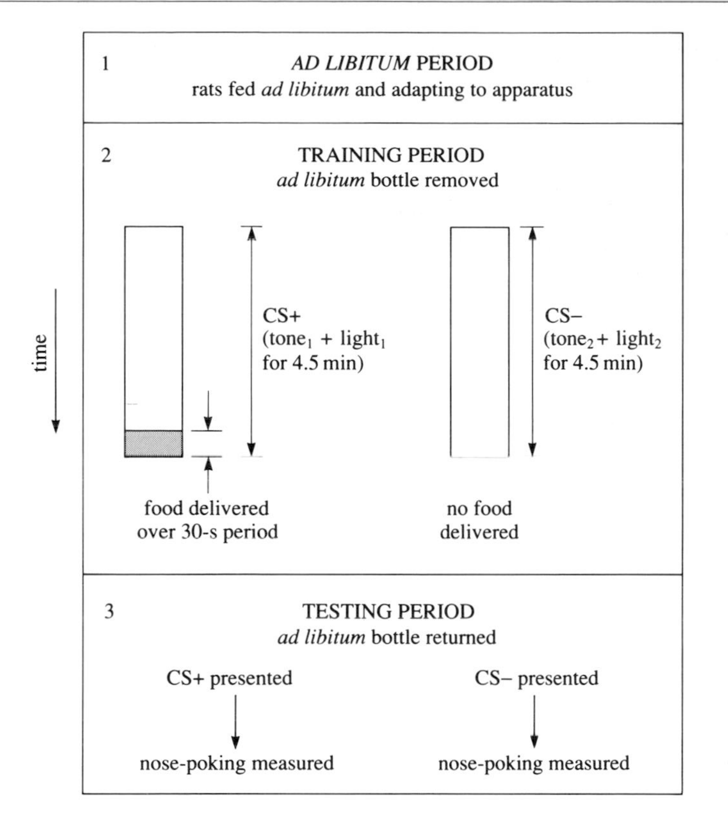

Figure 2.7 Summary of Weingarten's experiment. For explanation, see text.

During the training period, a combination of a different tone and light ($tone_2$ + $light_2$) was also presented to the same rats but this was at times when food was *not* available. It was termed the CS–, meaning a signal that food was not about to appear. During the first 4 minutes of the signalled 4.5-minute period, the number of investigations of the food cup, termed 'nose-pokes', was measured. Figure 2.8 compares the amount of nose-poking induced by CS+ and CS– during the 11 days of conditioning, called the training period.

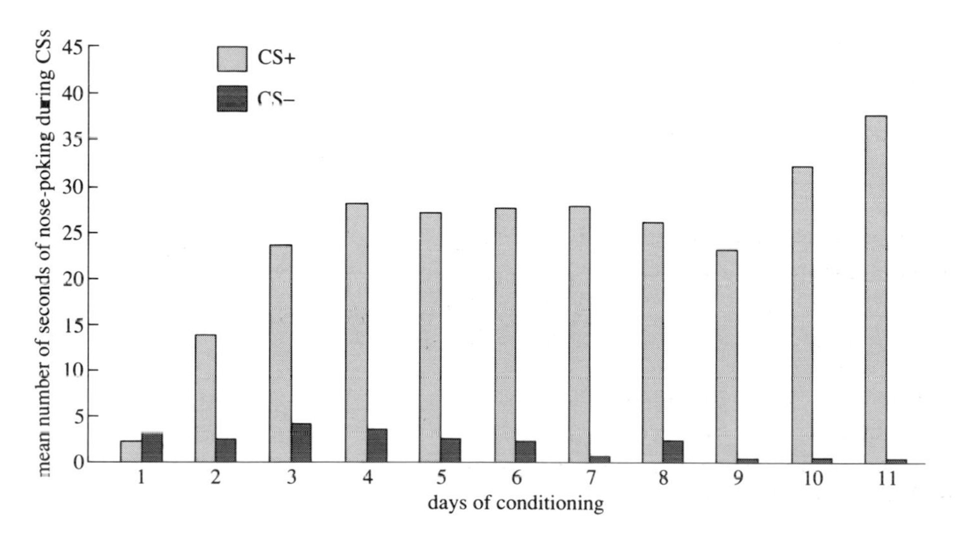

Figure 2.8 Mean time spent nose-poking into the food cup over the 11-day training period during CS+ and CS–.

□ Did the rats learn about the CS+ and the CS–?

■ Yes. There was an increase in the amount of nose-poking associated with the CS+, and a reduction to almost zero in the amount associated with the CS– over the training period.

Finally, for the testing period (3) the rats were placed on *ad libitum* feeding conditions, food being available in one of two places. A bottle containing liquid food was available throughout. At intervals, the identical liquid food was delivered into the food cup, signalled by a tone and light combination. There was a considerable amount of nose-poking induced by CS+ but none by CS–.

In further experiments, Weingarten looked at the effects of injecting a number of different substances on several aspects of feeding behaviour, e.g. nose-poking and amount ingested.

□ Considering Weingarten's experiment, what would constitute appetitive and consummatory behaviour?

■ Nose-poking is an example of appetitive behaviour. Ingestion of the food is an example of consummatory behaviour.

□ From the account of the role of CCK in feeding motivation, described in the last section, would you expect there to be differences in the effect of CCK injection on these two measures?

■ Yes. It was suggested that CCK is implicated in the process of terminating a meal. Therefore, whereas it was shown to affect the consummatory phase, it would not necessarily be expected to affect the appetitive phase.

Weingarten found that the amount of nose-poking and the latency to initiate feeding were not affected by injecting CCK. This suggests that the role of CCK is specifically one of terminating feeding, i.e. it affects the consummatory phase. By the measure used by Weingarten, it did not play a role in appetitive behaviour directed towards food.

2.2.4 The interaction of external and internal factors

Figure 2.3 showed both internal and external factors acting to determine feeding motivation. Further experiments have explored how they interact. A tone + food delivery to rats when hungry acquires the power to evoke feeding when they are satiated. However, does the tone act in complete isolation from the internal state? Would it be able to evoke feeding even in rats that are in energy surfeit? (A 'surfeit' means that the level of nutrients has been raised artificially to a value above that which the free-feeding animal would maintain.) Alternatively, would a signal from the body indicating surfeit cancel any excitation arising from the tone? Weingarten investigated the effect of placing food in the rat's stomach by a tube passed down the oesophagus (gullet). Fifteen minutes after this loading of the stomach, the rats were exposed to the tone. The loading resulted in a significant reduction in the size of meal eaten and an increase in the latency to eat (i.e. the time between the tone sounding and the rat eating). There was also a reduction in the amount of anticipatory nose-poking.

□ What control procedure would you suggest is necessary to conclude that a factor specific to the food load and feeding motivation was implicated?

■ Rats could be given a non-nutrient load. Suppose that the food (nutrient) load had a significant depressive effect relative to this control load. This would suggest that the lowered feeding was specific to the nutrient content of the load, rather than being due to stomach distension, for example.

Weingarten loaded the rat's stomach with a dilute saline solution as a control procedure. Relative to this control load, the nutrient load produced a significant reduction in both nose-poking and in the amount eaten. An increase in the latency to initiate eating was seen, implying that it was the nutrient content of the load that affected feeding.

Presentation of a CS+ is a powerful stimulus to ingestion in satiated rats. An interesting question is whether, over a 24-hour period, it actually induces them to eat more than they would normally. The alternative is that they cut down on how much they ingest at times other than when the CS+ is presented.

□ Figure 2.9 shows the result of an experiment that tested this. Interpret this result.

■ Total 24-hour intake is not significantly different, comparing a day on which the CS+ is presented and a day when it is not. The animals compensate for the feeding associated with the tone by eating less at other times.

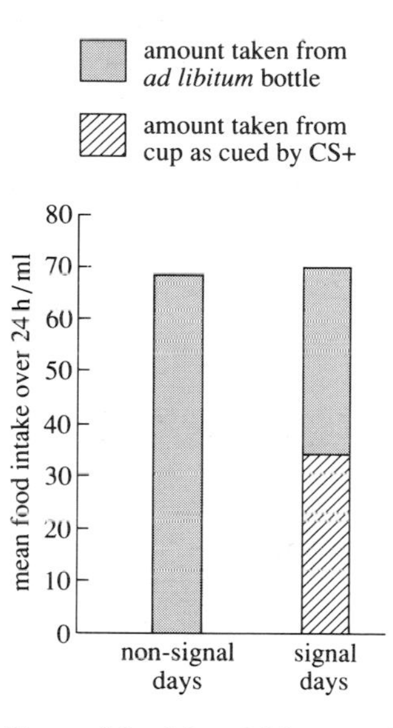

Figure 2.9 Mean 24-hour total food intake (i.e. from cup and *ad libitum* bottle) by rats on non-signal days (when CS+ was not presented) and signal days (presentations of CS+).

Thus, the experiments of Weingarten show clearly the role of a conditioned external factor acting in combination with internal factors in the determination of feeding motivation. The following section describes how further insight into the control of ingestion by both external and internal factors has been obtained.

2.2.5 The taste reactivity test

Section 7.2.1 in Book 1 described the change in pleasure rating of a stimulus as a function of the internal physiological state. Of course, one cannot in the same way ask a rat whether it finds a particular temperature stimulus or food pleasant or unpleasant! Normally, in the case of food intake, the best that researchers can hope to achieve in getting any comparable insight is to look at the variation in ingestion of a particular substance. It might be avidly ingested or ignored. Variations in ingestion can be studied following procedures that change either external or internal factors. For instance, depleting a rat's body of sodium will mean that relatively large amounts of a sodium chloride solution will be ingested. If the body is not depleted of sodium, typically little or none of the solution will be ingested.

Researchers have now devised a technique for looking more closely at how a rat processes sensory information, and how this processing varies with changes in internal or external factors. This ingenious method, which allows one, in effect, to ask a rat the question of how good does food taste, is termed the **taste reactivity test**.

In response to taste stimuli in the mouth, rats perform a number of stereotyped responses, as shown in Figure 2.10, *overleaf*. Some of these are positive in that they serve to mediate the ingestion of food, described as the ingestive sequence. Sucrose elicits the typical ingestion sequence, which consists of rhythmic mouth movements, tongue protrusions and paw licking. The other set of stereotyped responses are negative, meaning that they serve to minimize contact with the substance. This set is termed the aversion sequence, and consists of chin rubs, head shakes, and mouth wiping movements. A bitter substance such as quinine will elicit the typical aversion sequence.

Figure 2.10 Taste reactions of a rat to the application of solutions to its mouth. (a) Ingestion sequence elicited by a sucrose solution. (b) Aversion sequence, elicited by a quinine solution.

The taste reactivity test is carried out by means of implanting a tube in the rat's head, so as to allow the delivery of small quantities of solutions directly onto the tongue. In this way, the experimenter has control over the application of solutions to the rat's tongue. The rat is placed on a raised surface with a clear base, with an angled mirror below this; the reaction of the rat, seen in the mirror, is monitored by video camera.

This technique is highly sensitive to motivational variables. Changing the internal state of the rat has a profound effect upon ingestive responses. Sodium depletion turns the rat's reactions to a concentrated sodium chloride solution from negative to positive. Pairing a previously accepted flavour with nausea turns the reactions to the flavour from positive to negative. The rat acts towards the previously acceptable solution as if it now tastes bad. Thus, the taste reactivity test enables researchers to say more than just whether a substance is ingested or not, which is all that can be observed by measuring the animal's intake under normal conditions. The test shows that there can be more than one reaction to a substance, corresponding to it not being ingested: it might evoke either a neutral reaction or an aversive reaction. Motivation therefore has to do with rejection as well as acceptance.

Researchers paired a neutral stimulus (a tone) with the presentation of a sucrose solution to the tongue of a hungry rat. They then investigated the effect of presenting the tone at the same time as applying pure water to the tongue of the hungry rat. They already knew that pure water on its own elicited a response that consisted of both ingestive and aversive components.

□ Based on the results of Weingarten (just described) and an understanding of classical conditioning, what would you predict would be the response of the rat to the simultaneous presentation of a combination of tone and pure water?

■ There would be an increase in ingestive responses and a decrease in aversive responses.

This was indeed what was found. Owners of dogs or cats will probably not be altogether surprised at such results. The tenacity with which such animals follow the cues from the can opener must say something relevant about the roots of motivation. Although the laboratory context of presenting tones with the delivery of food might seem far removed from anything in the natural environment of the rat, it is important to note that wild rats have got where they are today by being opportunists. It is not difficult to see the adaptive significance of being able to form associations with cues that predict the availability of food. Furthermore, the fact that the cue can increase motivation means that, in some situations, food tends to be ingested as it becomes available. Such cues would also tend to provide persistence. Having engaged on a course of action, cues that in the past were paired with a particular goal will tend to keep the animal going to that goal.

Summary of Section 2.2

Feeding is determined by a number of factors, external and internal, acting in combination. There are detectors of the level of energy in the body, located in the liver and the brain, which play a role in motivating the animal to feed. When the level of glucose in the vicinity of these detectors falls, they change their rate of generating action potentials. Information from such receptors is integrated in the brain to give a command to seek food and ingest it.

The intrinsic quality of the available food also plays a role in feeding motivation. At a given internal energy state, the animal will often ingest some foods and decline others. In addition to intrinsic quality, in a wide variety of species, some foods that would otherwise be intrinsically acceptable are rejected because of earlier associations. In rats, food that has been ingested and followed by nausea will evoke an aversive reaction in a taste reactivity test.

Otherwise neutral cues that have been paired with the presentation of food can evoke ingestion in satiated rats. Such cues also change the rat's reaction to a given substance applied directly to the tongue.

The factors that switch feeding off are not simply a reversal of those that switch it on. For animals such as rodents that take relatively short, distinct meals, a meal will often be terminated before there is time for ingested nutrients to be absorbed in substantial amounts from the intestine. Amongst other factors, CCK is involved in a pathway that transmits signals from the gut to the brain and thereby plays a part in switching feeding off.

Section 2.2 looked at one motivational system: that underlying food intake. In studies of food intake in the laboratory, all conditions are normally held constant except for the manipulated independent variable, e.g. hours of food deprivation. The focus is upon one aspect of behaviour, and conditions are held so as to minimize 'distractions' from other motivational systems, e.g. sex. The environment is usually made deliberately simple. This is of course not what is found in the natural habitat. The next section moves slightly closer to the kind of situation an animal would confront in the wild. In the natural environment, an animal usually has the option to perform a number of different activities that can occupy its time.

2.3 Decision making and competition between activities

Section 2.2 concentrated upon a single motivational system, feeding. There has been a tradition within physiological psychology to look at such individual systems in isolation. However, insight into motivational systems has also been gained by investigations conducted in a broader context. That is to say, researchers have looked at factors outside the individual motivational system and considered how these factors affect it.

Engaging in one activity (e.g. incubation of eggs) is often incompatible with another activity (e.g. feeding, Book 1, Section 7.6). Performance of incubation would therefore involve inhibition being exerted upon other motivational systems. The present section returns to this broader approach to motivation and looks at other examples of inhibition. The first part of this section considers how feeding is influenced, not just by factors intrinsic to the feeding motivational system, but also by an interaction with the system of drinking motivation.

An animal will engage in one activity for a while and then switch to doing something else. The length of time that it spends performing a given activity is said to depend upon the strength of motivation underlying the activity. For example, imposing a period of water deprivation upon an animal means that, when water is available, it will spend relatively longer on drinking before switching to, say, grooming or feeding. In other words, animals could be said to be making *decisions* about what behaviour patterns to perform (Book 1, Section 7.1). The present section considers some of the factors that determine when switches between activities occur.

2.3.1 Feeding and drinking in rats

Figure 2.11 shows the amount of dry food eaten by rats over a 48-hour period, one group with, and the other group without, water.

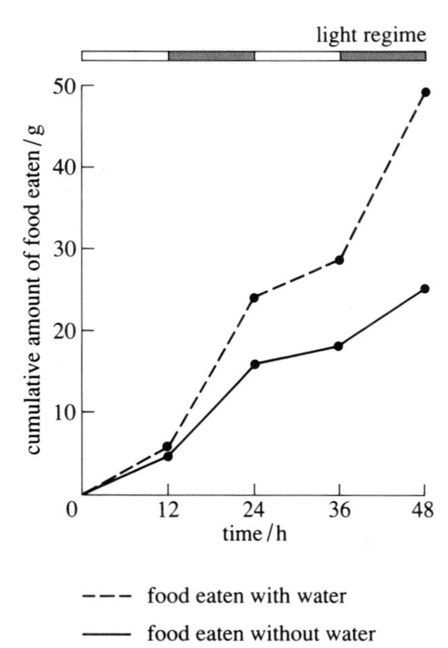

Figure 2.11 Cumulative amount of food eaten by rats over a 48-hour period, either with or without water available. Light periods were 0–12 and 24–36 hours (unshaded areas on the bar above the graph), and dark periods were 12–24 and 36–48 hours (shaded areas on the bar) from the start of the experiment.

□ Examine Figure 2.11 and describe two of the factors that control the intake of food.

■ (1) Rats having water available eat more than those not having water, and (2) there is a rhythm that affects intake. For both groups, intake is greater in the dark than in the light.

When their water intake is restricted, a variety of species reduce the amount of food that they eat. Broadly, there are two possible explanations for this. First, it might be that during a period of water deprivation, metabolism changes so that there is a lower nutrient requirement. Alternatively, it might be that in some way the intake of food is inhibited by drinking motivation. In fact, there is no evidence to suggest that metabolic processes are slowed up during water deprivation but as this section will show, there is evidence that dehydration and a high motivation to drink exert an inhibition upon feeding motivation.

The substance angiotensin has a powerful effect upon both drinking and the regulation of body fluids via the kidney. It is known to stimulate drinking directly; minute amounts introduced into the brain via a surgical tube are followed by copious drinking. To study the effects of angiotensin, rats were trained to press a lever in a Skinner box to obtain food pellets; water was not available during training sessions. Prior to the test, the rats had been deprived of food but not of water. Injections of angiotensin into the brain depressed lever-pressing for food reward.

This experiment is open to the interpretation that angiotensin acts directly upon feeding to inhibit it. Therefore, a further experiment was carried out to make sure that angiotensin was exerting its effect upon feeding via an induced increase in drinking motivation. To do this, a load of water was placed into the stomachs of a group of rats before injecting the animals with angiotensin.

□ What was the rationale for this?

■ Water in the stomach lowers drinking motivation (Book 1, Section 7.2.2). The assumption to be tested was that angiotensin exerts its effect upon feeding via an induced increase in drinking motivation. If this is so, water in the stomach might abolish or reduce the inhibitory effect on feeding.

This was exactly the effect obtained. It was concluded that angiotensin does not exert a direct effect upon feeding but does so via increased drinking motivation.

Evidence that there is an inhibitory effect of drinking motivation on feeding motivation also comes from results obtained by placing food-deprived rats in Skinner boxes where they had previously earned food but are now on extinction conditions.

□ What is meant by 'extinction conditions'?

■ Lever-presses no longer deliver food (Book 1, Section 6.3.4).

□ What would be the predicted effect of placing food-deprived rats on extinction conditions, comparing (a) animals with drinking motivation increased by injection of sodium chloride, with (b) animals injected with a neutral substance?

■ Rats in group (a) would be expected to press less frequently than those in group (b).

This is exactly what does happen.

What is the functional significance of the decrease in food intake shown by thirsty rats? Figure 2.12, *overleaf* was first presented in Book 1, Chapter 7. It shows one of the effects on the body of ingesting food.

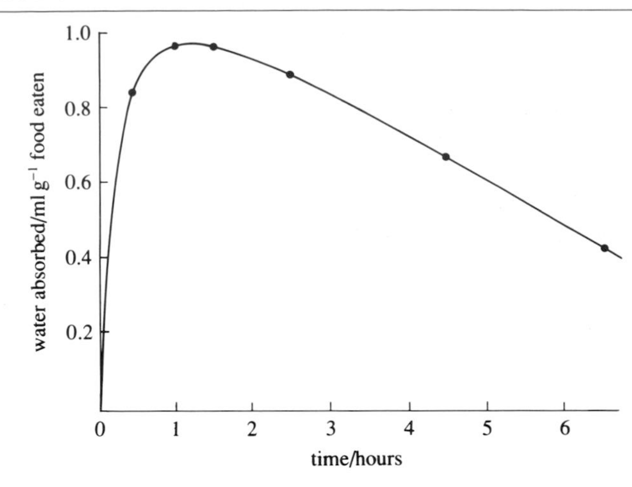

Figure 2.12 Graph showing water absorbed into the gut of a rat following a meal. Time refers to the time following the meal.

□ From looking at this result, what could be the functional significance of an inhibition of feeding motivation by drinking motivation?

■ The figure shows that food in the gut of an animal pulls water from the extracellular fluids into the gut. At a time of water deprivation, such a movement could be damaging to the circulation. Restricting food intake reduces this risk.

This section looked at how one motivational system involving ingestion can affect another. The following section also looks at an interaction between two systems, though two that are not concerned with ingestion. It shows the broader application of the principle of inhibition between systems.

2.3.2 Fear and pain

Defence, as described in Book 1, Section 7.1, has the properties of a motivational system. In response to fear, a variety of different behaviour patterns can be shown. Under some conditions, animals will freeze. At other times they will flee towards a place of safety. Book 3, Chapter 5 showed that pain can also serve a motivational role; the animal in pain is motivated to take action to reduce the noxious stimulation. However, the title of this sub-heading in the context of inhibition might surprise you. You might reason that there is no need to postulate any inhibition between fear and pain. One can be both frightened and in pain. Certainly, fear can be produced by association with pain. However, there is good experimental evidence that, under some conditions, fear can decrease pain. Most of the research has been carried out on rats and mice, though there is evidence that the phenomenon occurs in a range of species.

□ What is the adaptive value of defensive behaviour (Book 1, Chapter 7)?

■ Defence serves to protect the animal from intrusions into its environment. By fleeing it can save itself. Alternatively, it might show immobility. In this way it can often escape detection by a predator. In rodents, immobility serves to deflect an attack by a more dominant conspecific.

□ What is the adaptive value of pain (Book 3, Chapter 5)?

■ Pain causes the animal (a) in the short term, to favour injured tissue by, for example licking a wound, and (b) in the longer term, to rest and recuperate.

It is not difficult to imagine situations in which behaviour motivated by pain would be at odds with that motivated by fear. A defeated animal, or one showing immobility for whatever reason, might be better able to maintain immobility if it were less motivated to attend to its wounds. An animal not distracted or hampered by tissue damage would also be better able to flee from danger.

It is possible to make objective measurements of pain responsivity, by exposing animals to fear-evoking stimuli and seeing whether the threshold for reacting to a noxious stimulus changes. A large number of studies on various species and several different fear-evoking stimuli have shown unambiguously that the threshold is raised. In rats, simple exposure to the odour of a stressed conspecific raises the pain threshold. This phenomenon is termed **stress-induced analgesia**.

□ What is the meaning of the term analgesia?

■ Analgesia is a process that acts in opposition to pain. An analgesic substance is one that reduces the perception of pain (see Book 3, Chapter 5).

One of the most common ways of measuring analgesia is to employ the tail flick test (Book 3, Section 5.5.5). In this test, the rat's tail is placed over a lamp. The lamp is switched on and the latency, i.e. the time until the rat moves its tail away from the heat source, is measured. The longer the latency, the stronger the analgesia. In an earlier phase of the experiment, a tone is paired with an electric shock so that the tone acquires the capacity to elicit fear. The tone is sounded just before the tail flick test is given and it is reliably found that the latency to move the tail is increased.

□ In many cases, injection of the opiate-antagonist naloxone abolishes stress-induced analgesia. What is the implication of this?

■ This implies that the stress-induced analgesia is brought about by opioids, which would be released by the fear-evoking stimulus.

The last two sections have shown examples of the interaction between two systems. Though showing more of the complexity of real life than studies of individual systems, none the less in the natural environment an animal would normally be exposed to many competing demands. One of the best worked out studies of the complexity of interactions between systems has been carried out on an invertebrate, and forms the topic of the next section.

2.3.3 Competition in *Pleurobranchaea*

In the 1970s, William J. Davis of the University of California at Santa Cruz looked at the process of competition between activities in a carnivorous gastropod (snail) *Pleurobranchaea* (Davis, 1979).

Figure 2.13 represents a model of the animal's hierarchy of behavioural control. The behaviour termed 'righting' refers to the fact that if *Pleurobranchaea* is turned upside down it will tend to right itself. An arrow runs from a dominant to subdominant activity in the hierarchy. Where two activities appear to be at an equal level in the hierarchy, a double-headed arrow is shown running between them. For instance, feeding is shown to be dominant to righting. If the animal is turned upside down and food is then presented, it feeds and remains inverted for over an hour before it rights itself.

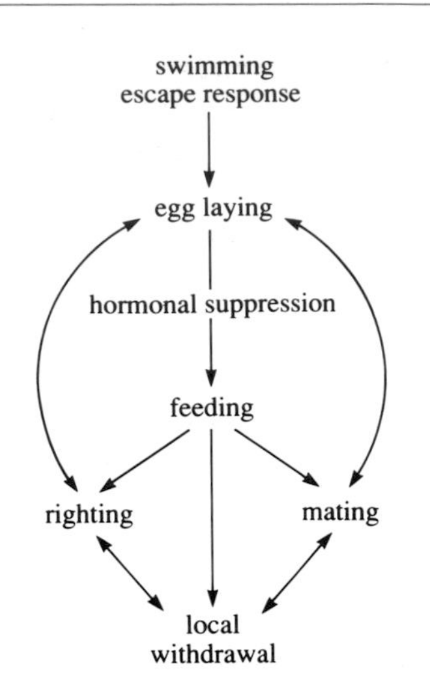

Figure 2.13 Behavioural hierarchy in *Pleurobranchaea*. See text for details.

After presentation of a tactile stimulus to the mouth region, the animal normally performs a withdrawal response which takes it away from the stimulus. However, if the same tactile stimulus is presented simultaneously with food, the withdrawal response is not shown. Something associated with food and feeding exerts inhibition on the withdrawal response, thereby preventing the animal from withdrawing from the tactile stimulus.

One might have supposed that the ability of food stimuli to inhibit righting depends upon the strength of motivation to feed. A satiated animal might show little inhibition. This was found not to be so; presentation of food even to a satiated animal, showing little or no tendency to ingest it, just as effectively inhibited righting as in a hungry one. Davis proposed that the chemoreceptors that detect food are the source of the signal that inhibits righting. By contrast, in satiated animals, presentation of food is no longer able to inhibit local withdrawal. Thus, one could describe the point of origin of the inhibitory signal upon local withdrawal as feeding motivation. This distinction between righting and withdrawal is shown in Figure 2.14. Davis was able to identify the neuron which carried information from the feeding system to the neurons responsible for local withdrawal. Electrical stimulation of this neuron when the animal was not feeding completely inhibited withdrawal.

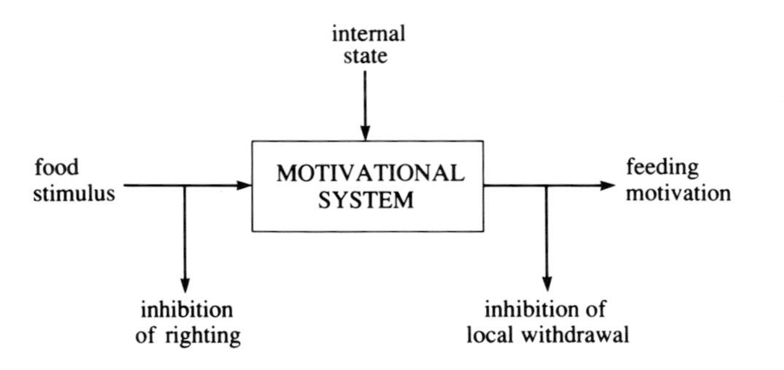

Figure 2.14 Model of the processes of inhibition in *Pleurobranchaea*. Detection of food inhibits righting. Feeding motivation depends upon the food stimulus and the internal state of the animal, and inhibits local withdrawal. In a state of satiety, presentation of food does not inhibit local withdrawal.

Feeding also takes precedence over mating. If a member of a pair of *Pleurobranchaea* is offered food during copulation, it will abandon copulation in order to eat.

☐ From functional considerations, what does the dominance of feeding over sex suggest?

■ That encounters with food are more rare than encounters with the opposite sex. Therefore, at least under the conditions observed, the animal's fitness is maximized by taking the opportunity to feed whenever food is available.

In general, if food supply is somewhat uncertain, it is often imperative to take food when it is available.

□ Is there a behaviour pattern higher in the hierarchy than feeding?

■ Figure 2.13 shows that egg laying takes priority over feeding. It exerts inhibition upon feeding.

□ What might be the functional significance of this?

■ Were there not to be such inhibition, the animal might eat its own eggs—not a good strategy to maximize fitness!

The inhibition from egg laying to feeding is hormonal. The same hormone that induces the animal to lay eggs is responsible for exerting inhibition upon feeding.

□ Is egg laying subordinate to any other activity?

■ It is subordinate to escape swimming. At a time of danger, escape must take priority over all other activities. Figure 2.13 therefore shows escape to be top of the hierarchy.

Further research will probably enable the details of these controls to be established. Already a lot is known about competition and hierarchies among activities in *Pleurobranchaea*, such as the roles of neural links and hormones in mediating inhibition from one system to another. To what extent *Pleurobranchaea* can serve as a model for the processes of competition in other species has yet to be established.

Summary of Section 2.3

Behaviour can be more fully understood by considering more than one system at a time and looking at the 'decisions' that animals make. The word 'decisions' is in inverted commas so as to emphasize that you should not think in terms of conscious decision making. Rather, all that is implied is that in the nervous system there exist processes that are influenced by factors underlying more than one motivation and, from the incoming information, animals are able to select one particular course of action. To achieve this implies that inhibition is being exerted on the systems underlying behaviour patterns that are not being expressed.

In a variety of species, drinking motivation inhibits feeding. The adaptive value of this is that food in the gut would pull water from the extracellular fluid into the gut, thus exacerbating dehydration. Another example given was of inhibition from fear to pain. A suggestion for the adaptive value of this is that a defeated animal would be better able either to maintain immobility or to flee if it did not attend to its wounds. In *Pleurobranchaea* inhibition is exerted from feeding to righting and from egg laying to feeding.

This section looked at principles that apply to more than one system, i.e. competition and inhibition. The following section also looks for general principles underlying behaviour, though it concentrates on feeding as an example. It looks at the processes within the nervous system that underlie behaviour. Systems of brain and behaviour appear intimidatingly complex and to get any kind of insight it is necessary to simplify them. Without such simplification, understanding them would prove to be impossible. Section 2.2 simplified by looking at an individual system. Section 2.4 also simplifies by looking at the role of one particular neurotransmitter in motivation.

2.4 Neurons, neurotransmitters and motivation

Any behaviour pattern involves the integrated activity of many brain regions and a multitude of transmitter substances all interacting with each other. However, to make a start at understanding a complex system it is sometimes necessary to simplify it. One approach has been to focus attention upon particular aspects of a given behaviour pattern and to ask what is the role of particular brain regions and neurotransmitters in producing this behaviour. To this end, the neurotransmitter, its agonists or antagonists (Book 2, Section 4.5), can be injected in particular brain regions. Alternatively, some researchers try to lesion the brain region to see what effect this has on behaviour. Suppose that injection of a neurotransmitter into a brain region were to increase greatly the amount of water that a rat drinks in 24 hours.

□ Give some possible explanations for this.

■ (1) The transmitter might have an effect on a brain region that underlies drinking motivation, increasing the level of motivation, (2) the transmitter might increase the rate of urination, in which case the effect upon drinking would be secondary, or (3) feeding might be increased and drinking would be secondary to this. You might think of still other possibilities.

□ How would you attempt to distinguish between (1) and (2)?

■ There are at least two experiments that you could do. You could observe the rate of urination following injection in the absence of available drinking water, to see whether it is abnormally high. You could study the rat's reaction to water in the mouth, using the taste reactivity test. If the effect is on motivational processes you might expect to see an increase in ingestive reactions. If the effect is directly upon urination you would not expect to see any immediate effect of the injection on taste reactivity.

Investigations that relate neurotransmitters to behaviour are fraught with difficulties. For example, suppose it is believed that a certain neurotransmitter is closely involved in the processes underlying feeding. To verify this, a particular drug, an antagonist to the neurotransmitter, is injected into a rat and it lowers food intake. It might be concluded that the antagonist has lowered feeding motivation and therefore this confirms the normal role of the neurotransmitter in exciting feeding. The dosage of antagonist injected is increased and food intake comes down still more. It is concluded that the drug has had a still more potent effect upon feeding motivation. This might be true, but without further evidence one cannot be sure. The action might be a more general one not specific to the feeding system. At increased doses, the drug might sedate the rat or block neuromuscular transmission. It might make the rat feel ill. It might do these things in addition to a specific lowering of feeding motivation. Alternatively, with increasing doses, there might have been no further lowering of the motivation to feed. Either way, lowered food intake is predicted. Cunning experimentation is called for to tease apart these explanations.

A transmitter that has been the focus of much investigation is dopamine, since it appears to play a crucial role in motivation, and it will be considered in the next section.

2.4.1 Mode of action of dopamine and the effects of antagonists

It is a well-known effect that, if an animal is trained in a Skinner box to obtain food or water, injection of dopamine antagonists interferes with the appetitive behaviour of lever-pressing.

□ Give some possible explanations for this.

■ (1) The drug might sedate the animal or make it feel ill, (2) it might impair its motor output, (3) it might impair its sensory–motor coordination, or (4) it might specifically lower its motivation to seek food.

So what does it take to show an unambiguous effect of dopamine antagonists acting on motivation? It is not necessary to show that factors 1–3 never apply: any or all of them might apply according to dose and site of injection. When injected into regions of the brain concerned with motor output, dopamine antagonists do indeed disrupt performance. To argue convincingly for an effect on motivation, it is necessary to find specific injection sites and demonstrate an effect on behaviour that would be predicted by the hypothesis that motivation has been lowered but which would not be predicted as a consequence of factors 1–3.

A number of pieces of evidence suggests that dopamine antagonists have an effect on motivation. It now appears that they have a rather specific disruptive effect. Suppose an animal is trained in an operant task: lever-pressing for the reward of food. It is then placed on extinction conditions. At first, frequency of lever-pressing increases to above normal, but it then declines. Over several trials, the frequency of lever-pressing can be brought to almost zero. Rats injected with dopamine antagonists, even when reinforced with food for lever-pressing, behave similarly to animals on extinction. The latency of injected rats to start lever-pressing in the Skinner box is normal. Frequency of lever-pressing is usually elevated at first (compared to the day prior to injection) but declines over several sessions. Similarly, rats trained in a maze at first show normal running speeds after injection, and only later do they slow up. Over a number of daily sessions, a decline in running speed is seen.

One powerful reward for normal rats is electrical self-stimulation of certain brain regions (Book 2, Section 10.5), which tend to be dopamine rich. Rats readily learn such a task and soon appear to be addicted to the habit. Brain stimulation is also a convenient means for looking at the action of dopamine antagonists. Figure 2.15 shows the behaviour of rats that had been trained for the reward of electrical brain stimulation. One group of rats is tested under extinction conditions.

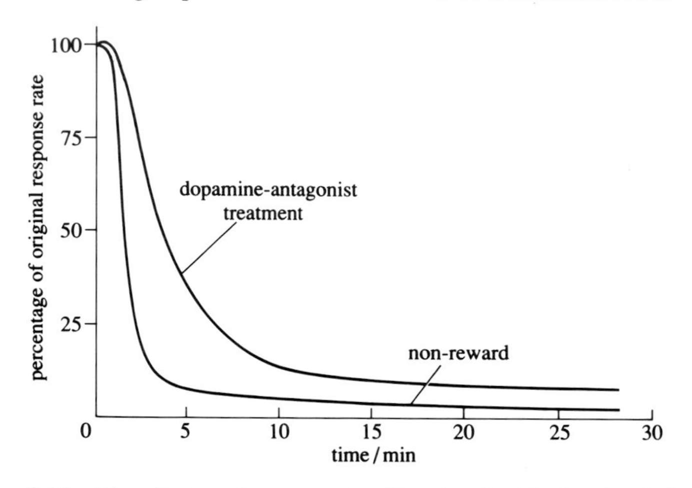

Figure 2.15 The effect on lever-pressing (for electrical brain stimulation) of imposing extinction conditions, i.e. non-reward or dopamine antagonist (pimozide) treatment.

□ What would extinction conditions consist of?

■ Electrical brain stimulation would no longer be obtained by lever-pressing.

Note the decline in responding when electrical stimulation no longer follows lever-pressing. Another group of rats was tested with electrical brain stimulation earned by lever-pressing, but under the effect of pimozide, a dopamine antagonist.

□ What is the effect of the drug on lever pressing?

■ The drug appears to have a similar effect to placing the animal on extinction conditions.

In other words, a combination of reward plus a dopamine antagonist becomes non-rewarding.

Just as a hungry rat can be taught a preference for the side of a T-maze in which it receives food, so can a rat be taught to prefer the side of a T-maze in which electrical stimulation of the brain is given. If, after a clear preference for the stimulation-associated side appears, dopamine antagonists are given and the rat is tested, on the following day the preference for the stimulation-associated side is sharply reduced. This again suggests that the drug reduces the *motivation* to go to the side on which reward was received.

Of relevance to the argument that dopamine antagonists have an effect on motivation is the phenomenon of **priming**. Priming is the revival of a particular behaviour performed as an operant task by presenting, under the control of the experimenter, the reward normally obtained in the task. A particular behaviour that is undergoing extinction as a result of omission of reward can often be revived in strength in this way. Suppose that a rat has been trained to press a lever in a Skinner box to earn the reward of electrical stimulation of the brain through an implanted electrode. Now, if extinction conditions are imposed, as Figure 2.15 shows, the rat will slow down and then stop pressing the lever.

Suppose that when the rat has clearly given up, a brief electrical stimulation of the brain is given through the electrode by the experimenter's action rather than by the rat's. Typically, the rat will instantly resume lever-pressing. Exactly the same result is seen if a rat is trained for food reward in a Skinner box, extinction conditions imposed until the rat has given up lever-pressing, and then a free pellet is given. Contemporary motivation theories emphasize the importance of the *presentation* of the appropriate stimulus in playing a role in triggering motivation. Priming revives the animal's motivation towards the operant task.

Priming presents a neat way of testing the effect of dopamine antagonists. The fact that the rat injected with dopamine antagonists acts like a rat under extinction conditions suggests that the drug has the effect of making something to do with lever-pressing or the gain of food lose its value. If dopamine antagonists have this effect, then presenting a stimulus when the animal is under the influence of such a drug might be expected to diminish its priming value, as reflected in future behaviour. An experiment along these lines is summarized in Figure 2.16.

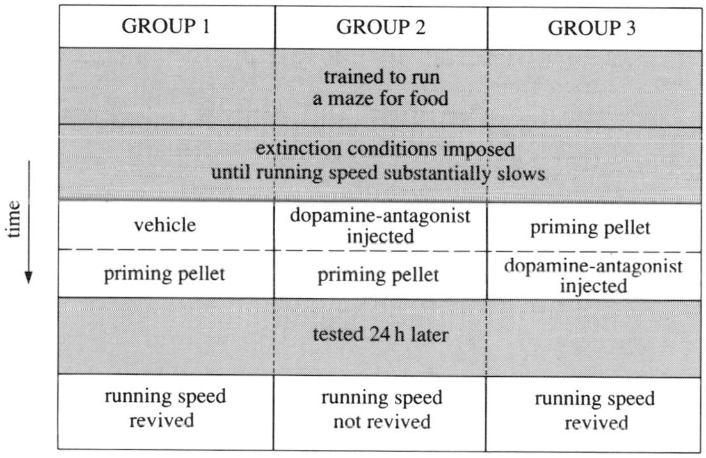

GROUP 1	GROUP 2	GROUP 3
	trained to run a maze for food	
	extinction conditions imposed until running speed substantially slows	
vehicle	dopamine-antagonist injected	priming pellet
priming pellet	priming pellet	dopamine-antagonist injected
	tested 24 h later	
running speed revived	running speed not revived	running speed revived

time

Figure 2.16 Summary of an experiment to test the effects of a dopamine antagonist on a priming stimulus. See text for details.

Three groups of rats are first trained in a runway to obtain food, and are then placed on extinction conditions until there is a substantial reduction in running speed. Group 1 are injected with a so-called 'vehicle' (an inactive substance) and they are given the opportunity to find a priming stimulus: on a single trial they find and ingest food in the goal box. Group 2 is injected with a dopamine antagonist, followed by the single priming pellet of food. For group 3, the dopamine antagonist was injected shortly *after* the priming food was eaten. Twenty-four hours later, the running speed of groups 1 and 3 is revived but not that of group 2. The motivational impact of the single pellet was lost by prior treatment with the dopamine antagonist.

□ What can be concluded by comparing the result of group 3 with the other two groups?

■ Group 3 revive their speed of running in spite of the dopamine antagonist. Therefore, the failure of group 2 to revive their running speed is not due to, say, fatigue or sedation induced by the drug. Rather it must be due to the interaction between the priming pellet of food and the drug. Prior treatment with the drug reduces the motivational impact of the priming stimulus.

At the doses tested, dopamine antagonists cause a particular kind of *motivational* change in the animal. The stimuli that impinge upon the animal and which play a part in its motivation appear to lose their motivational value, or at least have it lowered, if they are experienced while the animal is under the influence of a dopamine antagonist. This lowering outlives the presence of the drug in the body.

Dopamine antagonists exert an effect that is specific to the given environment experienced under the influence of the drug, e.g. a particular Skinner box and its history of associations for the animal. Place the same animal in a different context not experienced when under the effect of the drug and it will show normal operant performance.

□ What does this result say about the effect of the drug?

■ The animal is physically able to respond, but its motivation to do so is lowered in the context in which the drug was experienced. The change in the animal's responsiveness can be attributed to environmentally-specific motivational changes.

The motivational deficit is therefore an *interactive* one: it is the outcome of the interaction between the animal and the particular environmental context. This is distinct from sedation, fatigue and motor impairment as possible explanations for its behaviour. These are more general, environmentally-independent factors. However, one needs to be alert to the possibility of dopamine antagonists also disrupting behaviour by motor impairment or sedative action.

Summary of Section 2.4

Investigators sometimes inject neurotransmitters, their agonists or antagonists into the brain. Interpretation of the effects of such manipulations is fraught with difficulty. The substance might have an effect on behaviour but not in any simple way that casts any light upon the motivational systems that normally underlie expression of the behaviour observed. It might disrupt behaviour by inducing fatigue, sensory or motor impairment. One focus of investigation is the neurotransmitter dopamine. Dopamine antagonists cause changes in behaviour that are compatible with the effect of lowering motivation in a specific context.

The following section also considers dopamine and its role in motivation. It does so in the context of drug taking. It will be shown that an understanding of the principles of motivation is relevant to attempts to explain this behaviour.

2.5 Drug addiction

The use of heroin, cocaine, crack and other drugs is a topic seldom far from the media. The powerful motivational value of such drugs is evidenced by the craving that can arise from a history of association with them. In this section, some of the processes underlying this behaviour are described. Clearly, the factors involved in drug addiction are many and it is not possible to do justice to them all here. For instance, the section can do no more than acknowledge the importance of the kinds of crises faced by some young people taking up this habit, their social background, anxieties and depressions. These factors can set the scene for addiction. The emphasis here will be on what a knowledge of motivation and neurobiology can contribute, based largely upon animal experimentation.

2.5.1 A motivation model

Considerable insight into drug abuse has been gained by taking a motivation theory perspective and looking for features that the phenomenon has in common with the natural motivational systems of feeding, drinking and sex, as well as the not so natural motivation underlying electrical brain stimulation.

One obvious similarity is that rats will readily learn an operant task for the reward of an injection of intravenous opiates (e.g. heroin, morphine). The speed with which the habit is acquired in the absence of any measurable withdrawal effects strongly suggests that it is based upon the positive reinforcing effects of the drug, acting on the CNS. The properties of acquisition and extinction of an operant reinforced by opiates are closely comparable to those maintained by conventional reinforcers. Rats will lever-press for the reward of exceedingly small doses of opiates if they are applied to particular brain regions rich in dopaminergic neurons. If an opiate-antagonist is applied to the region, rats that had been trained to earn opiate infusions show an initial increase in responding and then a fall to a very low level.

□ What does this effect remind you of in terms of conventional rewards?

■ It is what happens when a food- or water-rewarded rat is placed under extinction conditions; initially, there is increased lever-pressing activity and then a decline in activity.

An important similarity between an operant task associated with opiate administration and one associated with conventional reinforcers is that they are both susceptible to priming (Section 2.4.1). If a rat learns an operant task rewarded with opiate injection into the brain and the opiates are then withheld, the response will extinguish. If a 'free' priming dose of opiate is then given by the experimenter, the rat will resume lever-pressing.

Priming is of central importance in theories of motivation in that it is an example of where motivation can be aroused by the *presence* of something, e.g. food or a drug in the brain. An understanding of the neural basis to priming is also of importance in the avoidance of relapse by addicts. After the response of a rat trained to obtain a drug has been extinguished, the habit can be reinstated not only by a 'free shot' of the substance used in training but also by other substances with similar effects on the CNS. For instance, heroin can reinstate cocaine delivery and vice versa. The serious implications of this for, say, a former heroin addict tempted occasionally to try cocaine are obvious.

As well as similarities with such conventional motivational systems as those underlying feeding and drinking, there are also some differences worth noting. The reinforcing impact of, say, a glucose solution for a food-deprived animal depends upon the stimulation of receptors (taste buds) in the mouth and the transmission of information along sensory neurons and then into the CNS.

□ In the case of glucose, what factor alters its reward value?

■ The internal state of the animal (see Section 2.2.5). The reward value would be less or even negative if the animal were to have been loaded with nutrients, as compared to being in a state of deprivation.

No comparable restraint exists in the case of drugs. Drugs exert their influence directly upon the CNS and so there are limited feedback effects of the type associated with, for example, feeding and drinking. This would appear to explain the potency of drugs relative to more conventional rewards like food. They do not require the mediation of sensory neurons and can presumably saturate receptors on CNS neurons that might never reach such levels of stimulation, even with the help of food, drink and sex. For this we must 'thank' the technology of a combination of the hypodermic syringe and the delivery of highly refined and synthesized forms of drug from far parts of the world. These drugs appear to act upon processes that would more normally be triggered by food or sexual stimuli. This means that the drug user can cheat and even overwhelm the normal processes of biological reinforcement.

Of course, another difference is that, unlike food or a mate, attraction towards drugs is the result of conditioning. There is nothing about opiates that is *intrinsically* attractive, nothing comparable, for example, to the sight and smell of a sexual partner. The attraction of opiates is critically dependent upon the consequences of experience with them. In this sense, the motivational properties of drugs are similar to that of electrical brain stimulation which also acts directly on the CNS. In neither case is there anything to taste, see or smell that has any motivational value, and there is no consummatory behaviour comparable to feeding or copulating. The route and means of taking drugs are various and of no intrinsic reinforcement value. However, the means *acquire* value by association with their consequences. For example, rats seek out locations associated in the past with either opiate infusion or electrical brain stimulation. Such associations are the topic of the next section.

2.5.2 Associative learning

In Weingarten's experiment (Section 2.2.3), it was shown that a neutral stimulus paired with food presentation can acquire the capacity to elicit feeding. There is evidence for a comparable effect in opiate administration, and it would appear to be of crucial social importance in gaining an understanding of relapse in drug addicts. Re-exposure to an environment in which drugs were taken in the past will often lead to relapse in drug-free individuals, even after extensive periods of abstinence. It suggests that stimuli associated in the past with drug-taking (originally 'neutral stimuli') become conditional stimuli that are able to arouse the urge to inject. In other words, conditional stimuli are able to elicit a motivational state much as a priming dose of a drug does.

Heroin addicts report being able to derive some pleasure from injecting inert substances. Some even report obtaining a drug-like euphoria when doing so in an environment similar to that associated with drug injection. Presumably, this forms the basis of the often-reported success of the pushers' con-trick in being able occasionally to adulterate drugs to the extent that the active ingredient is almost non-existent.

□ From your understanding of classical conditioning, what would be the predicted effect of the sustained injection of inert substances over a period of time?

■ The positive effect would wear off, i.e. extinction would take place. This does occur in practice.

To study the capacity of a neutral stimulus paired with drug administration to become a priming stimulus, the following experiment was carried out. Two groups of rats were trained to obtain intravenous cocaine by lever-pressing. For both groups, cocaine was delivered on a schedule in which, on average, every sixth lever press was rewarded. One group, termed the correlated group, was exposed to a tone on every reinforced response. The other group, termed the uncorrelated group, was exposed to the same number of tones but their presentation was not correlated with drug delivery (see Figure 2.17).

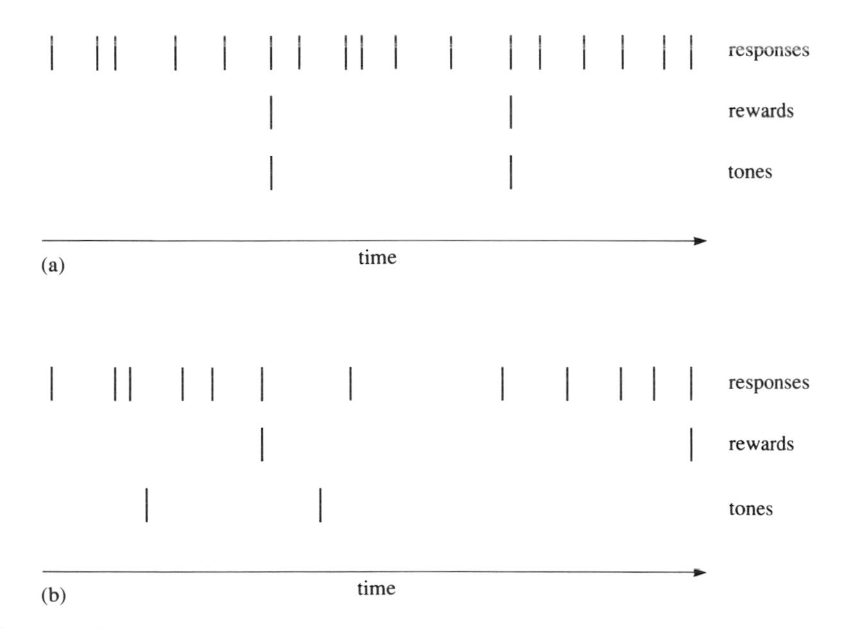

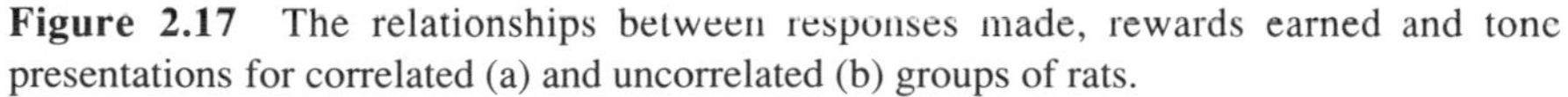

Figure 2.17 The relationships between responses made, rewards earned and tone presentations for correlated (a) and uncorrelated (b) groups of rats.

□ What was the reason for including the uncorrelated group?

■ There has to be a control group against which to compare the effect of the paired tone. One possibility would be to have used a group not exposed to a tone and compare their performance with those exposed to a tone. However, the researchers wanted to test the hypothesis that it was specifically the *pairing* of tone and drug that endowed the tone with its capacity to prime. By using an unpaired group, the history of both groups in terms of number of exposures to the tone was equated.

Both groups were then put under extinction conditions.

□ What would these conditions consist of?

■ Removing the contingency between lever-pressing and cocaine so that reward was not given for lever-pressing.

When the habit had extinguished, the tone was presented. For the correlated group but not the uncorrelated group, the tone reinstated lever-pressing, demonstrating that pairing the tone with drug delivery endowed it with the the capacity to elicit lever-pressing.

2.5.3 Positive and negative reinforcement aspects

Finally, you might be wondering why the chapter has emphasized the positive motivational aspects of opiates and cocaine rather than the aversive withdrawal effects.

□ In the terminology of learning theory (Book 1, Chapter 6), distinguish between drug-taking motivated by withdrawal symptoms and that motivated by obtaining euphoria.

■ Escape from withdrawal is an example of negative reinforcement. Obtaining a 'high' is an example of positive reinforcement.

Surely, you might feel, addicts, as normally portrayed, are locked in a fruitless attempt to escape the pain of withdrawal rather than to gain the euphoria of a 'high'. Clearly, painful withdrawal effects sometimes play an important role. Indeed, under some conditions, they can form the base for classical conditioning. Addicts sometimes get conditioned withdrawal symptoms on being exposed to an environment in which withdrawal symptoms were experienced in the past. However, the evidence from both humans and rats suggests that, for opiates, the positive reinforcement can sometimes be *sufficient* to explain much of what is happening.

Such an argument is not to deny that, in some cases, a negative reinforcement component is also present. Neither does it deny that addicts experience enormous stresses in their lives nor deny them the right to care and concern. This view has powerful clinical implications for pharmacological approaches to addiction: treating the withdrawal symptoms associated with detoxification will be insufficient. As Roy A. Wise from Concordia University in Quebec noted in 1988:

> ...it is a conclusion that since it goes against such ingrained and widely held assumptions must be reiterated each time it arises. Cravings based on the positive reinforcing effects of a drug will vary in strength with the memories for past reinforcements; these memories, in turn, will vary with the presence of secondary reinforcing stimuli or drug-associated incentives in the environment. They will remain long after any temporary discomfort associated with the physiological rebound effects that define physiological dependence syndromes...

A survey of addicts questioned about their reasons for relapse after a period of abstinence found that less than a third experienced conditioned withdrawal effects. Those experiencing them did so on rare occasions. The desire to obtain euphoria was the most common explanation given for relapse.

Summary of Section 2.5

Understanding of the behaviour of taking opiates and cocaine has been advanced by viewing this behaviour in the context of a knowledge of motivation. In spite of some obvious differences, the motivational system underlying this behaviour has similarities to other motivational systems. Animals will learn an operant task for the reward of opiates. The profile of acquisition and extinction of a task reinforced with opiates or cocaine is similar to that of tasks reinforced with food or water. Rats will acquire the habit of lever-pressing for the reward of minute doses of drugs to dopamine-rich areas of the brain. Following extinction of an operant task rewarded with such drugs, lever-pressing can be revived by a priming injection.

Classical conditioning plays a powerful role; neutral stimuli paired with the administration of drugs can acquire the capacity to motivate drug-taking.

Summary of Chapter 2

The emphasis of Chapter 2 has been on the role of both internal and external factors in determining behaviour. Although the primary focus has been on feeding, the principles discussed are more general ones. Other motivational systems (e.g. sex) are also dependent upon internal and external factors.

One of the internal factors in feeding is the energy state of the body as detected by, for example, glucoceptive neurons. That motivation towards food is increased by a fall in the internal energy level, of course, makes sense in adaptive terms as does the increase in drinking motivation caused by a fall in body fluids. It is the essential feature of homeostasis. In such terms, the results of the taste reactivity test also make sense. Ingestive responses are promoted in appropriate physiological states. By 'appropriate' what is meant is that ingestion changes the physiological state in an adaptive direction (e.g. low-energy states promote ingestion of food). Aversive reactions are promoted in physiological states that would be adversely affected by ingestion (e.g. the animal with a surplus of sodium in the body shows an aversive reaction to salt). Pairing a particular solution with nausea is followed by an aversive reaction to the substance. This is true even in physiological states that would, in the absence of the experience of nausea, promote ingestion of the substance. This result can be understood in terms of a broader concept of homeostasis. A substance that is associated with nausea, even if energy-bearing, is best avoided in the interests of defending overall physical well-being.

External factors that determine ingestion include the intrinsic properties of available foods and cues that have in the past been associated with food. Weingarten's procedure demonstrates the efficacy of conditional stimuli and the taste reactivity test demonstrates that they can affect the processing of sensory information on taste. A neutral chemical stimulus of pure water is rated more positively by being presented simultaneously with cues that have been paired with sucrose. What is the possible adaptive significance of such classical conditioning? In the examples given here it is possible to see where classical conditioning could give *persistence* to behaviour. Suppose an animal is foraging for the contents of nuts. The shell has been associated in the past with the nutritious nut. By increasing the animal's motivation level, contact with the shell of the nut will tend to keep the animal going.

□ This should remind you of a feedback process introduced in Book 1, Sections 7.2.2 and 7.6. What is it?

■ Positive feedback. The first contact of food in the mouth *increases* feeding motivation and gives persistence to behaviour in the face of competing demands.

Section 2.3.1 showed that animals do not always maintain one particular physiological variable constant at all times. There can be conflicting demands upon an animal. At times, homeostasis of one system must be partly sacrificed in the interests of another. For example, the water-deprived animal cuts down on its intake of food. A lizard will refrain from venturing onto the surface when to do so risks hyperthermia (overheating) (Book 1, Section 7.6). In so doing, the energy state falls but this is in the interests of overall physiological well-being of the body. Similarly, survival could be best served by a process of analgesia that comes into effect at times of fear. In this way, the response to tissue damage is suspended in the overall protection of the animal.

Finally, the chapter argued that drug-taking can be understood within a broad framework of explanation provided by looking at conventional motivational systems. Drugs such as heroin and cocaine tap into the reward processes that normally underlie conventional systems (e.g. sex and feeding). Therefore, it is not surprising that drug-taking shows some features in common with behaviour directed towards conventional rewards. For example, Section 2.4 described priming in both feeding and electrical stimulation of the brain, and Section 2.5.1 described it for drug administration. Section 2.2.3 described the effect that cues associated with food delivery have upon the feeding system, and Section 2.5.2 described the potency of cues paired with drug administration to arouse drug-taking. The fact that dopamine seems to be closely involved in the processes underlying drug-taking provides a further link between abnormal and normal motivations.

Section 2.2 looked at a single system (the shaded area was the feeding motivational system in Figure 2.1). Understanding can require consideration of more than one system; mutual interactions also need to be looked at (as represented in Figure 2.2). From a few examples you have seen how some of the processes underlying motivation can be identified in terms of the role of particular neurons, hormones and neurotransmitters. Throughout the discussion, caution has been emphasized. There are some profound problems of interpretation of the evidence. For example, the fact that injecting a certain substance might lower feeding does not tell us that the natural role of the substance is to induce satiety.

Objectives for Chapter 2

When you have completed this chapter, you should be able to:

2.1 Define and use, or recognize definitions and applications of, each of the terms printed in **bold** in the text.

2.2 Distinguish between appetitive and consummatory behaviour. (*Questions 2.1 and 2.2*)

2.3 Give an account of the processes involved in initiating and terminating feeding in rats, explaining the role of external and internal factors. (*Questions 2.2–2.4 and 2.8*)

2.4 Describe the taste reactivity test, and explain its relevance to understanding the determinants of ingestion. (*Question 2.5*)

2.5 Give an example of an interaction between motivational systems, explaining (a) the suggested causal basis of the interaction, and (b) its possible functional significance. (*Questions 2.6 and 2.7*)

2.6 Explain in what way caution is needed in interpreting data on the effects of injected substances on behaviour. (*Question 2.8*)

2.7 Describe how experimenters have tried to understand the possible actions of dopamine antagonists on motivation. (*Question 2.9*)

2.8 Give an account of drug-taking (e.g. opiates, cocaine) that relates it to other motivations and to underlying motivational processes. (*Questions 2.9 and 2.10*)

Questions for Chapter 2

Question 2.1 (*Objective 2.2*)
Which of the following are examples of appetitive behaviour and which are examples of consummatory behaviour?

(a) A thirsty animal is drinking.

(b) An animal is pressing the lever in a Skinner box for the reward of water.

(c) Two animals are copulating.

(d) A hungry animal is running along a maze towards the reward of food in the goal box.

Question 2.2 (*Objective 2.3*)
(a) What is the adaptive significance of short-term feedback loops in the feeding motivational system?
(b) What might be expected if they were to be rendered inoperative?

Question 2.3 (*Objective 2.3*)
In Weingarten's experiment, how would you attempt to eliminate the capacity of the CS+ to elicit feeding in satiated rats?

Question 2.4 (*Objectives 2.2 and 2.3*)
Extrapolating from Weingarten's result, what would be the expected effect of CCK injections on the rate that a rat presses a lever during extinction for a task previously reinforced with food.

Question 2.5 (*Objective 2.4*)
Which one of the following procedures would be expected to change a rat's reaction towards sodium chloride from negative to positive?

(a) The rat is deprived of water for a period of time.

(b) The rat is injected with concentrated sodium chloride.

(c) The rat is placed on a diet lacking sodium chloride.

Question 2.6 (*Objective 2.5*)
Some experimenters wished to test the hypothesis that fear induces analgesia. A tone was assumed to have acquired a fear-evoking capacity by being paired with electric shock in an early phase of the experiment. Then, using the tail flick test as a measure of analgesia, the tone was presented to induce analgesia.

(a) What would form a suitable control condition for this experiment?

(b) How could a possible role of opiates be implicated?

Question 2.7 (*Objective 2.5*)
With reference to Figure 2.13, what happens if food is presented to *Pleurobranchaea* during egg laying?

Question 2.8 (*Objectives 2.3 and 2.6*)
The following is part of an imaginary conversation between two SD206 students. Where would some caution be in order?

Bill I know serotonin is implicated in feeding motivation because injecting a serotonin-antagonist reduces feeding to almost nothing.

Mary I am sure the antagonist simply makes them feel sick and therefore serotonin has nothing to do with feeding motivation. Also, if serotonin plays the role you suggest, injecting a serotonin-agonist in rats would make them eat much more than normal and I know it doesn't have that effect. Jack tried injecting one at the end of a period of 24 hours of food deprivation. In the next 30 minutes when they had food again they ate no more than uninjected controls.

Question 2.9 (*Objectives 2.7 and 2.8*)
How might the impact of dopamine antagonists on motivation be examined with the help of (a) a rat that has been trained to lever-press for the reward of intravenous morphine, and (b) the phenomenon of priming? What results would be expected on the assumption that dopamine antagonists lower motivation to obtain the drug?

Question 2.10 (*Objective 2.8*)
Consider the experiment described in Section 2.5.2, which investigated classical conditioning, and make one hypothetical change to it. Rather than uncorrelated presentations of tone, as shown in Figure 2.17b, the control group receive no presentations of the tone prior to testing for its efficacy. Testing for efficacy of the tone is exactly as described. What would be the possible implications of such a design for the conclusions that could be drawn?

(a) Suppose in the test session, the tone was sounded and the control group rats resumed lever-pressing as frequently as the experimental group, exposed to pairings of drug and the tone. How could such a result be interpreted?

(b) Conversely, suppose that the control group did not resume lever pressing but the experimental group did. How would that alter the conclusions that could be drawn?

References

Davis, W. J. (1979) Behavioural hierarchies, *Trends in Neurosciences*, **2**, pp. 5–7.

Weingarten, H. P. (1984) Meal initiation controlled by learned cues: basic behavioural properties, *Appetite*, **5**, pp. 147–158.

Further reading

Ettenberg, A. (1989) Dopamine, neuroleptics and reinforced behaviour, *Neuroscience and Biobehavioural Reviews*, **13**, pp. 105–111.

Stewart, J., de Wit, H. and Eikelboom, R. (1984) Role of unconditioned and conditioned drug effects in the self-administration of opiates and stimulants, *Psychological Review*, **91**, pp. 251–268.

Toates, F. (1986) *Motivational Systems*, Cambridge University Press.

Wise, R. A. (1988) The neurobiology of craving: implications for the understanding and treatment of addiction, *Journal of Abnormal Psychology*, **97**, pp. 118–132.

York, D. A. (1990) Metabolic regulation of food intake, *Nutrition Reviews*, **48**, pp. 64–70.

CHAPTER 3
BIOLOGICAL CLOCKS

3.1 Introduction

Rhythms are a ubiquitous feature of living systems. These rhythms vary greatly in the time taken to complete a cycle. For rodents, activity normally shows a 24-hour cycle corresponding to the rotation of the earth (Chapter 1). Within this cycle, some animals are active in the light and others in the dark. Other cycles that have been discussed include the 4–5 day oestrous cycle of female rats (Book 2, Section 10.3.2). There are also cycles of behaviour that reflect the tilting of the earth on its axis and take one year to go through a cycle. This chapter looks at a number of such cycles, and asks what their external and internal determinants are.

Biological cycles or rhythms pervade every aspect of physiology and behaviour. Chapter 1 gave examples of rhythms that are intrinsic to the organism (e.g. feeding in rodents); they continue even in complete light or darkness. Daily rhythms that have been shown to be intrinsic (endogenous) are called **circadian** (from the Latin 'circa' and 'dies', meaning 'about a day'). Within the cell, there are 24-hour rhythms in biochemical processes. Cells of a similar kind are found in the same tissue, so there are 24-hour rhythms at the level of tissue, such as those of the endocrine glands; tissues make up organs so that organs also show 24-hour rhythms. Thus, animals' biochemistry, physiology and endocrinology are not simply the constant processes often naively assumed: they fluctuate in cycles.

All these rhythms are regulated so that they keep a proper relationship to each other as well as to the external light/dark (LD) cycle; they have to peak and trough at the correct time of day for their respective functions. Therefore, it is necessary to entertain the idea of a biological clock that regulates these internal rhythms. Finally, the multiplicity of rhythms means that consideration of time of day has important consequences for some areas of biology and medicine. Thus, in any organism, there is a complex organization of daily rhythms, a kind of 'temporal homeostasis' essential for health. The potency of drug action can vary with time of day, due to the basic biology of organisms showing daily rhythmicity, as can the magnitude of unwanted side-effects. Therefore, optimal timing for therapeutic regimes is an important application of the principles discussed in this chapter; despite this, the general medical practitioner is still in the habit of prescribing medication to be administered no more precisely than 'three times a day'!

3.1.1 Terminology of rhythms

In Figure 3.1a, the terms used to describe a rhythm are illustrated: this example might be the path traced by the pendulum of a clock. The term *period* describes the time taken for the cycle to complete itself; an obvious example of a period would be the time taken for the pendulum to go back to its starting point after one cycle.

- □ What is the period of the human menstrual cycle?
- ■ Approximately 28 days.

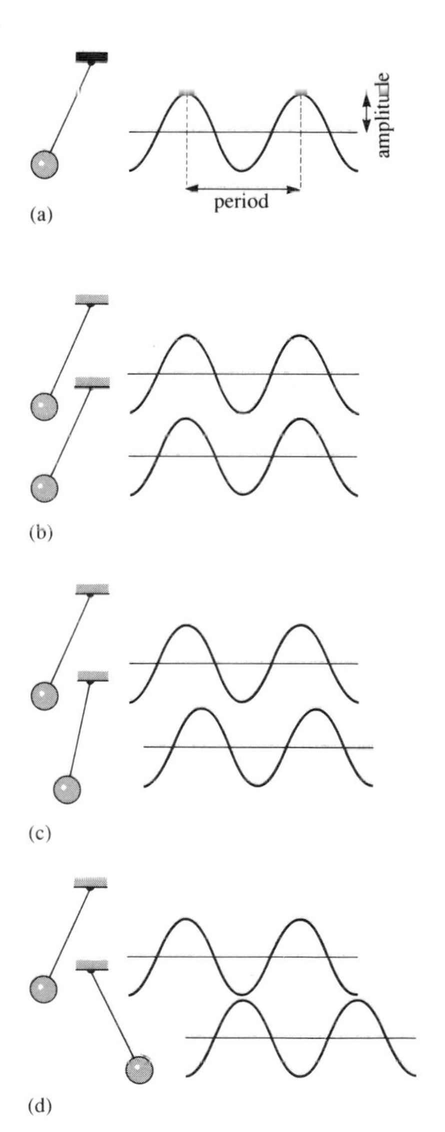

Figure 3.1 (a) Terms used to describe a cycle. (b) Two rhythms in phase. (c) Two rhythms slightly out of phase, and (d) two rhythms 180° out of phase.

The extent of the rhythm, i.e. the maximum departure from the average of the peak and trough, is called the *amplitude*. Each point on a cycle is a *phase*; the peak or trough of the rhythm are examples of phases. Figure 3.1b illustrates the path taken by two pendulums whose movements are exactly synchronized. The pendulums would be described as being *in phase*. When two rhythms do not coincide in their timing, they are said to be *out of phase* (shown in Figure 3.1c). In Figure 3.1d the rhythms are still further out of phase: when one rhythm peaks the other is at its trough and vice-versa. The amount that two rhythms are out of phase is measured in degrees—the two rhythms in Figure 3.1d would be said to be 180° out of phase.

Figure 3.2 relates this concept to the kind of rhythm found in behaviour. If the activity in part (a) was said to be in phase with the light/dark cycle, that shown in (b) would be said to have shifted out of phase with the light/dark cycle.

□ Has the period of the rhythm changed between (a) and (b)?

■ No. It is 24 hours in both cases. Only the phase relationship has changed.

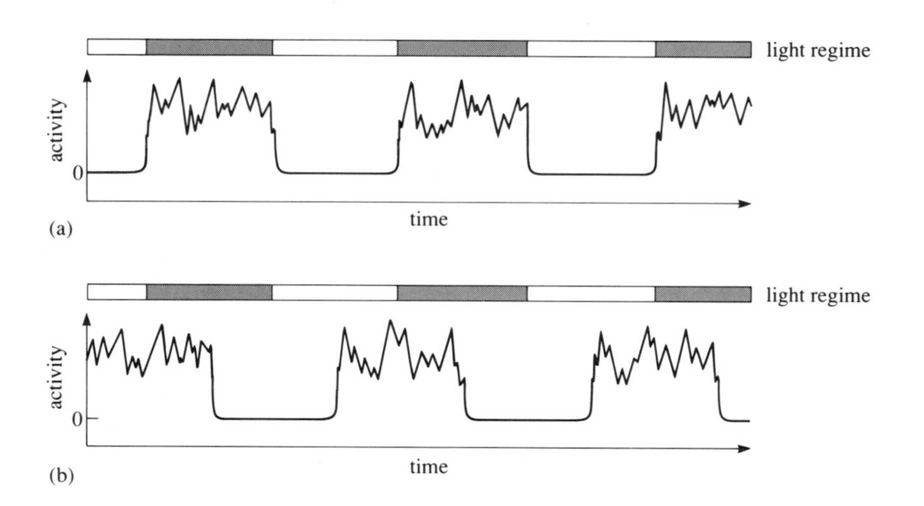

Figure 3.2 The relationship between activity and the 24-hour light/dark cycle. In the bar representing the light regime, the shaded areas indicate periods of darkness (dark phases), and the unshaded areas periods of light (light phases). (a) Activity is in phase with the light/dark cycle. (b) Activity is out of phase with the light/dark cycle.

So much for the basic terminology of how to describe rhythms—the next two sections look at some examples of biological rhythms in different species.

3.2 Examples of circadian rhythms

3.2.1 Activity in the cockroach

Rhythmic changes in activity can be monitored in a number of different species by recording locomotor activity, often measured as the amount of running in a wheel. A very precise onset to activity is frequently seen, with the end of activity often being more variable. Records of wheel-running data are usually presented in the form of what is termed an *actogram*, which is explained with reference to Figure 3.3. The figure shows a running wheel and the running wheel activity of a cockroach. When the cockroach is active the pen makes one upward mark on the trace for every turn of the wheel. When the wheel is turning rapidly without stopping, the trace becomes solid.

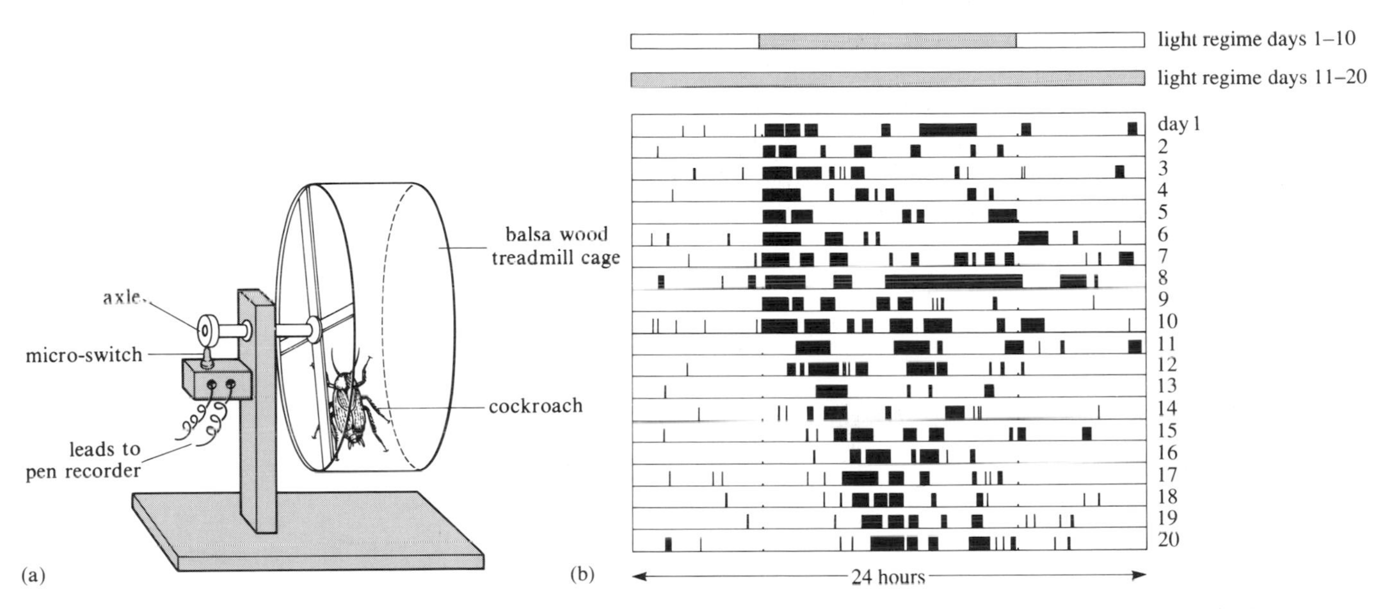

Figure 3.3 Rhythms in a cockroach. (a) Apparatus for recording locomotor activity. As the cockroach runs or walks, the wheel rotates, triggering a microswitch connected to a pen recorder each time the wheel completes a revolution. (b) Activity record (actogram) for a cockroach kept for 20 days in the running wheel. The light regime was 12 hours light/12 hours dark for days 1–10. For days 11–20, the regime was one of continuous dark. In the bars representing the light regime, the unshaded areas indicate light phases, and the shaded areas dark phases.

During days 1–10, the cockroach is exposed to a light/dark cycle with a period of 24 hours. During days 11–20, it is subjected to continuous darkness.

□ Is there evidence of a circadian rhythm in Figure 3.3b?

■ Yes. The animal is more active in the dark phase of the 24 hours, *and* a rhythm persists when it is placed in continuous dark.

□ Would the rhythm be described as endogenous or exogenous?

■ It is endogenous (generated within), a true circadian rhythm.

□ Using the original light/dark cycle as reference, does activity shift out of phase when the animal is exposed to continuous dark?

■ Yes. The animal's active period is progressively delayed.

When an endogenous rhythm is in synchrony with the light/dark cycle over a number of cycles, the light/dark cycle is said to **entrain** the rhythm of the animal.

Similar results are found if the behaviour of rodents is examined, as the next section shows.

3.2.2 Activity in rats

Researchers often find it convenient to examine data on rhythmic activity with the help of what is termed a *double plot* consisting of two columns, each of 24 hours width (see Figure 3.4). Each 24-hour period consists of so many hours of light and so many hours of dark, usually determined by the experimenter. In Figure 3.4, there are 12 hours of light and 12 hours of dark, as shown by the bar representing the light regime at the top of the figure. The top row of the activity record (in the column to the right) is the pattern of an animal during the first 24 hours of observation (1). In the next row, in the column to the left, the record of the first 24 hours (1) is duplicated and continues into the record of the second 24-hour period of observation (2). By plotting data in this way each 48 hours, made up of two 24-hour periods, can be seen easily. This means that each kind of transition, light to dark and dark to light, can be viewed with ease.

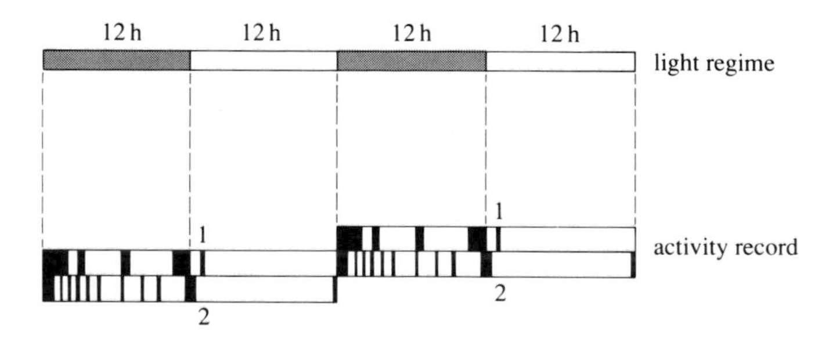

Figure 3.4 Double plot. For explanation, see text. In the bar representing the light regime, shaded areas indicate dark phases and unshaded areas light phases.

Plotted in this way, Figure 3.5 shows the running wheel activity rhythm of an adult female rat.

In the first stage of Figure 3.5, activity was monitored under a light/dark (LD) cycle. Since the animal is **nocturnal** (active during the night), by definition most activity occurs in the dark. (This contrasts with humans, for example, who are said to be **diurnal** because they are normally active during the day.) Every fourth or fifth night there is a greater incidence of activity. Female rats have a 4–5 day oestrous cycle and, on the late afternoon to evening of pro-oestrus (that part of the oestrous cycle just before the period—oestrus—in which they are sexually receptive), they become very active. Typically, females from this strain of rats run in their wheels only the equivalent of 0.25 km per night on ordinary nights but on the night of oestrus, 4–5 km may be run.

□ During exposure to the light/dark cycle, would the activity cycle of the rat shown in Figure 3.5 be said to *entrained* to the light/dark cycle?

■ Yes, because the rhythm is synchronized with the environmental light/dark cycle.

In the second stage of Figure 3.5, the light/dark cycle is stopped and the laboratory placed under continuous lighting (referred to as LL). The absence of the light/dark cycle eliminates the major entraining agent, termed the **zeitgeber** (German for 'time-giver', pronounced roughly as 'zyte gay-burr'), so that in the absence of this environmental time-giver, the internal clock takes on its own intrinsic periodicity. This internal timing device underlying rhythmicity is called a **biological clock**. Normally, the biological clock is reset each day to 24 hours by the lights going on and off.

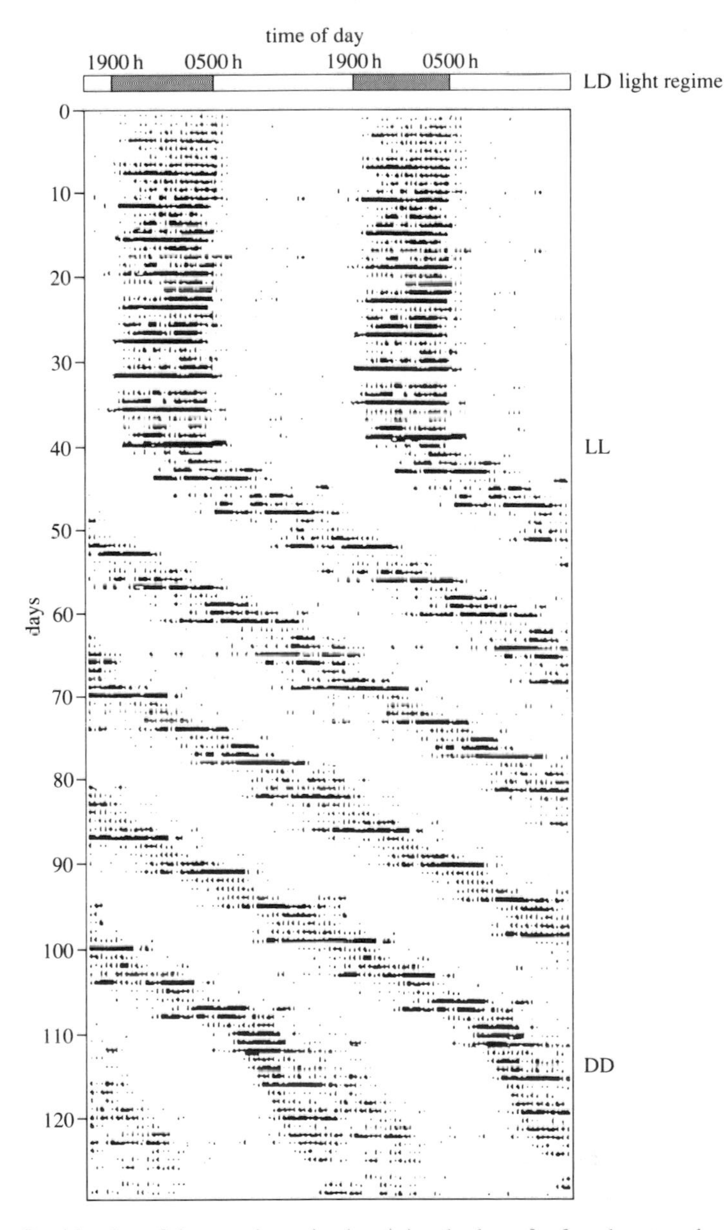

Figure 3.5 Double plot of the running wheel activity rhythm of a female rat under three different lighting conditions: first, a normal light/dark cycle, LD (light 14 hours: dark 10 hours per day for days 1–39), then constant light, LL (days 40–113), followed by constant darkness, DD (days 114–130). Two identical plots are placed side-by-side, with the right-hand plot 1 day higher than the left-hand plot for ease of comparison between successive days. In the bar representing the LD light regime, shaded areas indicate dark phases and unshaded areas light phases.

For the strain of rat shown in Figure 3.5 (but not for all rats), the clock's endogenous period is greater than 24 hours. As you can see, this results in the onset of activity occurring later and later each 24 hours. The rhythm is now said to be **free-running**, i.e. running in the absence of an apparent zeitgeber, and the period of the free-running activity rhythm (termed τ, pronounced roughly as 'tow', to rhyme with 'cow') is taken as being indicative of the period of the biological clock.

Any circadian cycle of activity is divided into two phases, not necessarily of equal duration; the active phase and the inactive phase, when the animal is at rest and often asleep. The relationship between these phases under the free-running state may be different from that under entrained conditions. If the rhythm's period lengthens under constant conditions, then either the active or inactive phases, or both, must have lengthened.

□ When the conditions change from LL to DD, what happens to τ?

■ It shortens.

□ Is the period under DD longer or shorter than under LD conditions?

■ It is slightly longer, hence the gradual shift to the right over several days.

A distinction needs to be drawn between the period of (a) the biological clock itself (which, as will be shown later, is in the CNS), (b) the output from the clock (which in this case is locomotor activity), and (c) the coupling process between the clock and the output. Generally, it is assumed that the period of the output variable that is measured under free-running conditions (the 'hands of the clock') reflects exactly the period of the clock itself. However, since the latter cannot be studied directly, there is no way of knowing whether this assumption is correct.

In Figure 3.5, the persistence of the cyclic changes in behaviour at intervals of approximately, but not exactly, 24 hours, for long periods of time, in the absence of zeitgebers, indicates that this is a *circadian rhythm*. What is important in this definition is the *persistence of the rhythm in the (apparent) absence of time cues*; animals have many daily rhythms, but not all of these have been shown to be circadian. If rhythmicity is lost under constant lighting conditions the rhythm is not deemed to be circadian. Rather, it must be a passive system that is driven to oscillate by changes in the external environment, i.e. it is exogenous. Circadian rhythms are active systems capable of self-sustained oscillations, i.e. they are endogenous. The classic test to discriminate between exogenous and endogenous rhythms is, therefore, to remove all cyclic environmental agents and determine whether the rhythm still persists.

3.2.3 Activity in humans

In 1962 the French cave researcher Michel Siffre camped in a tent in an underground cavern in total isolation for 62 days. This cavern was 114 m under the Alps, with the ambient temperature being kept reasonably constant at 0 °C. The cavern was in total darkness except for a battery-powered light, so that Siffre had no indication of the changes in light and darkness above ground associated with the day/night cycle. Furthermore, Siffre had no watch, no clock and no radio, his only contact with the external world being by field telephone. Each time Siffre went to bed, awoke, ate a meal, etc., he telephoned the surface so that these times could be recorded by his collaborators, Siffre himself having no indication of objective time. Thus, as far as was technically possible, all zeitgebers were eliminated. However, researchers can never be absolutely certain that this has been completely achieved. For example, in principle at least, the rotation of the earth might induce a magnetic change that could provide a cue to time.

Above ground, the length of one cycle of activity (the period of the cycle) for the French population was, of course, 24 hours. What happened to the sleep/wake cycle of Siffre? He no longer lived a 24-hour day, but instead his day lengthened so that, on average, it was 30 minutes longer. Thus, his sleep/wake cycle fell increasingly out of phase with that of the day/night cycle and the activity cycle of the population on the surface.

Figure 3.6 shows the result. The horizontal axis indicates the time of day, while the vertical axis measures number of days; successive days are stacked underneath one another. As can be seen, on the first few days, Siffre awoke at approximately 0800–1000 h in the morning, and went to bed at approximately 2000–2200 h in the evening.

- □ What is the value of τ, the period of his internal clock, revealed under free-running conditions?
- ■ Approximately 24.5 hours (so that on average, the onset of his waking and sleeping shifts by 30 minutes per day).

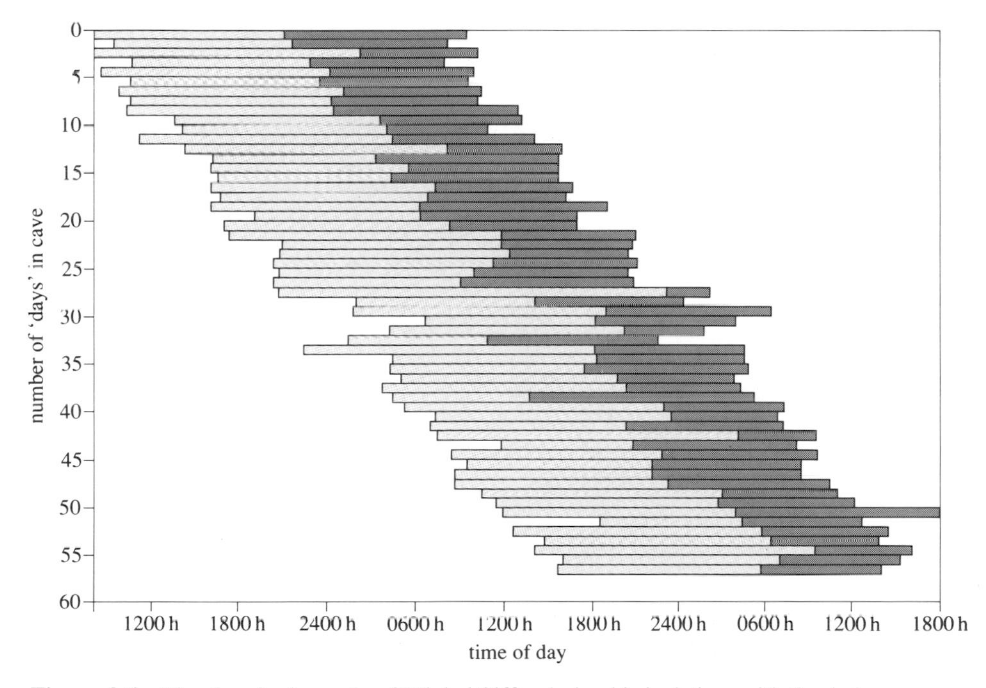

Figure 3.6 The sleep/wake cycle of Michel Siffre during his isolation, with the dark grey areas indicating sleeping/resting periods, and the light grey areas awake/active periods. Note that the width of the plot covers a period of 57 hours so that the free-running rhythm can be seen more clearly.

By day 25, Siffre was waking up at approximately 2000–2100 h in the evening and going to bed at approximately 0800 h in the morning. Siffre was now 12 hours *out of phase* with the world above ground. By days 40–45, Siffre had again returned in phase to the external world, getting up at around 0700–0800 h and going to bed at about 2200 h.

There are a number of interesting features of Siffre's underground sojourn. First, he underestimated the passage of time markedly; he thought it was 20 August when the experiment finished, when it was, in fact, 14 September. His subjective calculation of time was underestimated by 25 days. From his diary, we know that often what was recorded as a short nap, a siesta, was in fact an 8-hour sleep. Because he underestimated sleep length, he therefore underestimated his time spent underground. Siffre's behaviour also indicates that the timing and duration of human sleep is greatly influenced by a biological clock and is not just the reflection of some social or cognitive factor.

□ Does the evidence from Siffre show that in humans the biological clock is endogenous or exogenous?

■ Endogenous. The fundamental timing information comes from some internal time keeper and not from the exogenous daily cycles of light and dark, temperature and humidity etc, because these were all kept constant in the cave.

Siffre's mood appeared to vary with the phase of his internal cycle in relation to the external cycle above ground of day and night. His daily log indicated that his mood was pessimistic, except for 4 days at the time when his sleep/wake cycle again came into phase with the external world (in the period of days 43–48). This raises the interesting question of whether biological timing could have an influence on a person's mood and mental health.

After Siffre's experiment, there were some ten further cave studies over the next decade. The most extensive and informative human studies were carried out in an underground bunker at the Max Planck Institute at Andechs in southern Germany. All the experiments produced the same basic findings: (1) the human sleep/wake cycle is influenced by an internal clock whose period is usually longer than 24 hours (on average 24.5 hours for a normal healthy individual), and (2) all aspects of the internal environment, as well as behaviour and performance, are influenced by this clock.

Human sleep/wake cycles, such as that shown in Figure 3.6, are very variable in terms of sleep onset, duration and ending: humans can 'choose' to remain awake (e.g. when reading a good book) even if their biological clock is telling them they are tired enough to go to sleep. In contrast, many mammals exhibit rest/activity cycles under constant conditions that are tremendously precise, with an accuracy within a few minutes each day. This accuracy allows the researcher to observe small changes in daily cycles as a result of their experimental manipulations. In addition, the facilities for study and the number of human volunteers for such long-term experiments are obviously limited, so investigators have to rely on animal research for much basic information.

Summary of Section 3.2

In a variety of species, including cockroaches, rats and humans, the endogenous nature of the biological clock is revealed in conditions of constant illumination (free-running conditions), where the rhythm persists. The period (τ) of the rhythm under these conditions typically is slightly different from 24 hours. When the animal is exposed to a 24-hour light/dark cycle, rhythms, such as those of activity, show a 24-hour period and the rhythm entrains to the light/dark cycle. The entraining agent, which is most commonly light, is termed the zeitgeber (time-giver). A circadian rhythm is one that (a) free-runs in conditions of constant illumination, and (b) has a τ of about 24 hours, such that it entrains to the 24-hour light/dark cycle. To give a more complete picture of the properties of biological rhythms, the following section looks briefly at rhythms of a period other than 24 hours.

3.3 Infradian and ultradian rhythms

Rhythms that cycle less often than circadian rhythms are known as *infradian rhythms*.

□ What does this say about the period of the infradian rhythms?

■ They have a period longer than 24 hours.

The oestrous cycle modulation of locomotive activity seen in Figure 3.5 is an example of an infradian rhythm. Another example would be the 28-day human menstrual cycle. An important set of infradian rhythms have a frequency of about one year, corresponding to the second important geophysical zeitgeber.

□ What could the zeitgeber consist of for such rhythms?

■ Variation over the year of the ratio of light and dark within the 24-hour period.

The amount of daylight, which under natural conditions varies with the seasons in temperate and polar regions, is known as the *photoperiod*.

Cycles that have a period shorter than circadian (i.e. cycle more often) are known as *ultradian*. An example is the 90–120 minute cycle of rapid eye movement (REM) sleep in humans, or the episodic release of certain hormones which is superimposed upon a circadian rhythm. An example of an ultradian rhythm is that of growth hormone release in rats (Figure 3.7, *overleaf*), as measured by blood sampling at intervals of a few minutes.

□ What is the approximate period of the rhythm underlying growth hormone release in rats?

■ The figure shows that the rhythm in growth hormone release has a period of just over 3 hours.

This periodicity is apparently unrelated to laboratory light cycles, stages of sleep and activity, and seems to be independent of other hormonal changes and metabolic disturbances.

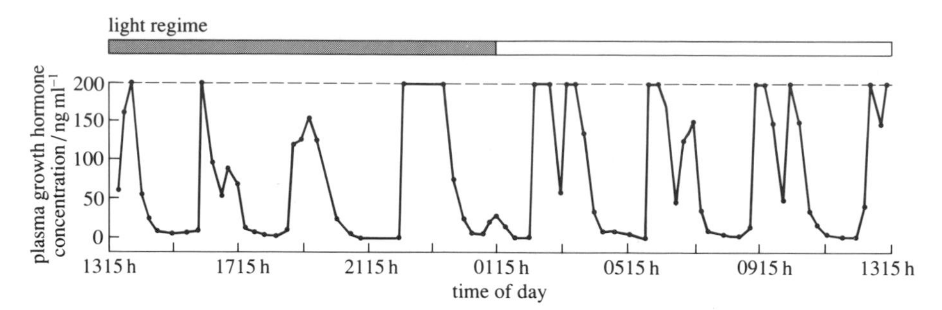

Figure 3.7 Changes in plasma growth hormone concentration over a 24-hour period in a rat. Blood is sampled every 12 minutes via a permanently implanted surgical tube which allows the rat to move freely around its cage. The dotted line indicates the maximum measurement that could be taken. In the bar representing the light regime the shaded area indicates the dark phase, and the unshaded area the light phase.

Later sections will consider infradian and ultradian rhythms again. However, most of the interest in rhythms has been in the circadian type and this forms the bulk of Chapter 3, to which the discussion now returns.

3.4 Some properties of circadian rhythms

3.4.1 Re-entrainment

Suppose the timing of the environmental zeitgeber is changed. What happens to a rhythm that was previously entrained to the zeitgeber? Figure 3.8 shows the result of an experiment designed to find out. As Figure 3.3b showed, the cockroach is most active in the dark phase. Figure 3.8 also shows this effect; on day 1 activity begins at the time of lights out. On day 2 the timing of the light/dark cycle is changed by the experimenter.

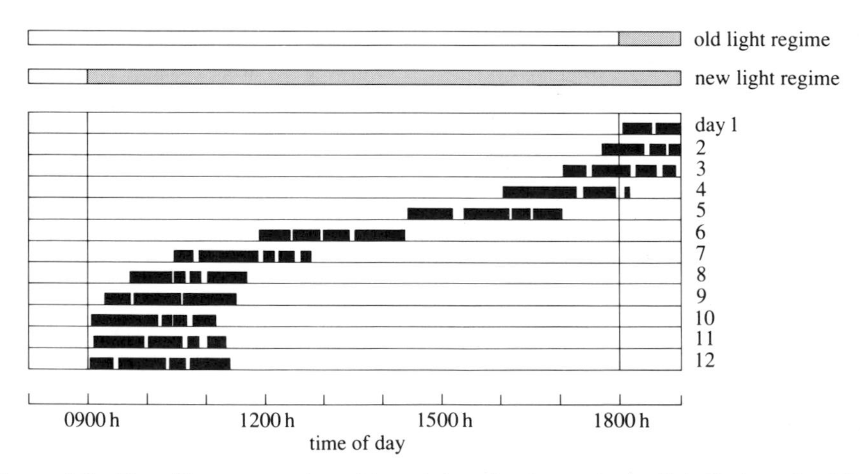

Figure 3.8 The effect on a cockroach's activity of a change in the light/dark cycle. After day 1, the light/dark cycle was changed, so that lights out occurred 9 hours earlier. In the bars representing the light regimes, the shaded areas indicate dark phases, and the unshaded areas light phases.

- □ Interpret the effect of the change on the cockroach's activity.
- ■ The activity rhythm gradually shifts so as to come back into phase with the zeitgeber. This takes about 10 days to complete.

- □ This might remind you of something you have experienced or have at least read about. What is it?
- ■ Jet lag. In this case, as with the cockroach, it takes time to adjust to a time-zone shift.

The process of coming back into phase with the zeitgeber is termed **re-entrainment**.

Similarly, if an environmental light/dark cycle had been introduced during the LL or DD periods in Figure 3.5, the animal would re-entrain, i.e. the phase would reset. For this to happen, the period of the light/dark cycle has to be nearly equal to that of τ. This resetting usually takes a number of cycles and is achieved by a number of small phase shifts.

3.4.2 Determinants of τ

Consider again the third stage of Figure 3.5, where the laboratory lighting conditions are changed from constant light (LL, days 40–113) to constant darkness (DD, days 114–130). The locomotor rhythm still persists but shortens in its period; it is now closer to 24 hours. This illustrates an important relationship between the intensity of light and τ. for nocturnal mammals, τ increases with increasing light intensity, whereas for diurnal mammals it decreases.

τ is generally considered to be both temperature independent and chemically independent. Today, it is thought that these points may not be strictly true since circadian rhythms of some mammals under certain conditions can be manipulated by temperature cycles and the administration of certain chemicals or hormones. However, the wisdom was based upon sound logic. If animals have a clock measuring and giving out time to the body, the cells of the clock must be specially organized or protected in some way, compared with other cells. Most cells of the body speed up or slow down their biochemical reactions with temperature changes, and are heavily influenced by the chemistry of the internal environment. However, if the cells of the clock were to react to temperature or chemical changes in the same way that ordinary cells do, it would not be possible to keep biological time accurately. Timing would be erratic. Therefore, in some way not yet understood, the cells of the clock must compensate for body temperature changes to a large extent.

3.4.3 Individual differences in τ

The τ for any individual in a group of rats or other mammals in any one experiment will be different from other rats, even though the difference may be small. Since the circadian periods of individuals are different, after many days of free-running, the individual rhythms will be out of phase with each other. This has important consequences that can lead to misinterpretation of results. Suppose, for example, that one wishes to study a particular hormone. This can be done only by repeated blood sampling of each individual rat. Suppose that blood is obtained at, say, 4-hour intervals. Examine Figure 3.9. Under light/dark entrained conditions (Figure 3.9a) all animals are in phase.

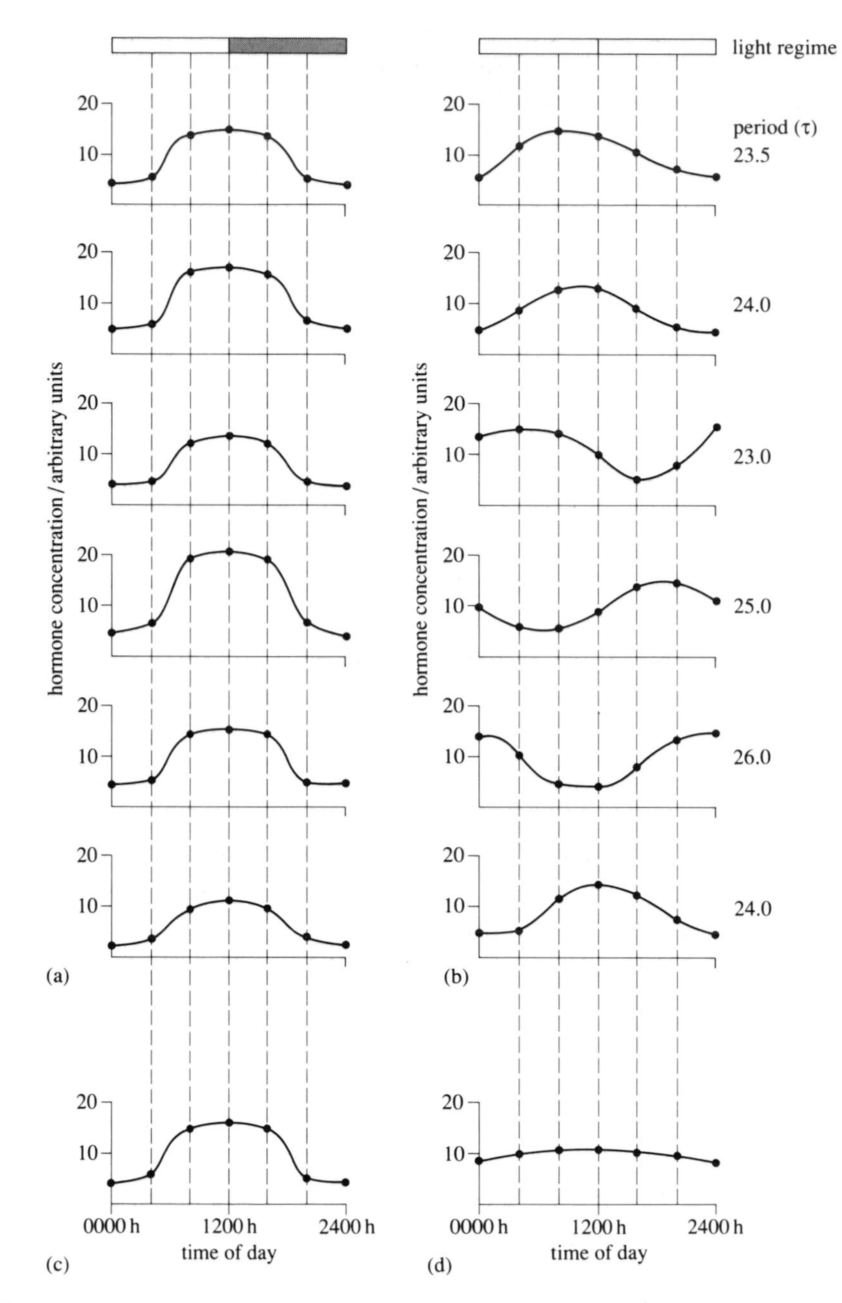

Figure 3.9 Hormone concentration sampled every 4 hours throughout the light/dark cycle. (a) shows results from six rats entrained to a 24-hour period by a light regime of 12 hours light/12 hours dark (LD), and (b) shows results from six rats free-running in constant light (LL). (c) represents the group mean data for (a), and (d) the group mean data for (b). In the bars representing the light regimes, the shaded area indicates a dark phase, and the unshaded areas light phases.

□ How does the mean result for the group in Figure 3.9c compare with the result for any given individual?

■ The mean is rather close to what each individual is doing. An impression of the rhythm in each individual can be gained by looking at the mean.

Thus, if plasma is sampled at the same clock time, e.g. every 4 hours, rhythmicity is observed. Now compare the entrained condition with the free-running condition shown in Figure 3.9b.

□ How does the mean result for the group in Figure 3.9d compare with the result for each individual animal?

■ The mean shows no evidence of rhythmicity but it would be quite wrong to conclude from this that no individual exhibits rhythmicity, since they all do.

Under free-running conditions (Figure 3.9b), all animals are at different phases because the periods of their free-running rhythms are different. When plasma is sampled every 4 hours, for any one time point some animals will be peaking and some will be at their trough. When these plasma values are averaged, the group mean will have eliminated all the individual rhythmicity. Ignorance of the free-running phenomena of circadian rhythms has erroneously resulted in reports in the literature of non-rhythmicity for certain variables because of this averaging effect on phase.

Looking at the mean result, the conclusion might be wrongly drawn that, since no rhythm exists under constant conditions, any rhythmicity under light/dark conditions must be exogenous and hence not circadian. To avoid this problem, looked at under these conditions, animals have to be sampled, not at a fixed, objective clock time, but at equivalent times in their rhythms, i.e. circadian time. A convenient way to do this is to use the start of the active period of running in the wheel as an easily observable marker, a frame of reference.

In sampling human blood or in timing injections of a drug, a knowledge of the underlying rhythm could be crucial. In treating patients who might have undergone a shift in a rhythm due to, for example, travel from one time-zone to another, a knowledge of the individual rather than the group norm could prove invaluable.

Summary of Section 3.4

If the timing of the light/dark zeitgeber is changed, re-entrainment will take place gradually. Under conditions of constant light or dark, τ varies as a function of the lighting condition. For nocturnal animals, it increases with a change from dark to light. The converse is true of diurnal mammals. There are individual differences in τ, such that if a number of animals are left free-running, they will get out of phase. The next section looks at some of the properties of circadian rhythms and asks how entrainment occurs in the natural environment.

3.5 Understanding the circadian clock—a black box approach

Under natural conditions of light/dark, how is the system entrained every day to 24 hours? We know that the light/dark cycle is the dominant zeitgeber of mammals, but how exactly does entrainment take place? How does light synchronize the circadian clock? What are the characteristics of this clock?

The characteristics of the system in rodents have been examined selectively by its response characteristics to a series of environmental manipulations, particularly of lighting schedules. This approach has allowed researchers to deal with a wealth of experimentally derived data and make testable predictions about the control mechanism without reverting to brain manipulations. This 'black box' approach contrasts with the neurobiological approach consisting of, for example, lesions, and chemical and electrical stimulation, which are discussed later. In this black-box approach, neither the anatomical location nor the biochemical systems of the postulated clock have been investigated directly, but the researchers have still gained a profound insight into the properties of the mechanisms subserving the circadian system.

They found that the characteristics of light in terms of its colour, intensity and duration are relatively unimportant in determining the clock's response to the light. The key factor is *when* the light is on, i.e. there are 'critical periods' when the clock responds to light stimuli in a characteristic manner.

3.5.1 The phase-response curve

The basic experimental method involved rodents which were housed individually in continuous darkness, each with access to a running wheel. The following discussion of light-induced phase shifts in activity is based on data from such nocturnal rodents. Species studied were golden hamsters (*Mesocricetus auratus*), white-footed mice (*Peromyscus leucopus*), deermice (*Peromyscus* sp.) and house mice (*Mus musculus*): these were usually males, to avoid complications from the oestrous cycle in females. All these species are nocturnal in their activity; thus they avoid the bright light of day. However, as will be shown later, the basic concepts of light sensitivity seem to apply to all living things other than bacteria and viruses.

As was discussed earlier, reintroducing a light/dark zeitgeber to an animal maintained under constant LL or DD conditions results in a progressive move in the animal's behavioural rhythm so that it becomes entrained to the new zeitgeber. However, the phase of a rhythm can also be shifted simply by pulses of light applied at particular times during the behavioural cycle. The results of these experiments reveal important properties about the nature of the endogenous rhythm underlying behaviour.

Individual animals are first allowed to free-run in DD until they have a stable baseline period. Then, once every 2 weeks, a single light pulse is administered for a duration of 1 hour (but 15 minutes is adequate). The effect of the light pulse in terms of a phase shift of the timing of the onset of the activity rhythm is observed. For a given animal, the light pulse is applied at different times relative to the onset of the activity rhythm, and the magnitude and direction of the phase-shift induced is then measured. A graph of the phase shift obtained against the phase in the rhythm at which the light pulse is given is called a **phase-response curve (PRC)** to light.

This method is illustrated in Figure 3.10. The frame of reference is defined by the animal's activity. In this case, the phase when it is active is defined as the *subjective night* and the phase when it is inactive is termed the *subjective day*. (For a free-running rhythm, these will be out of phase for most of the time with objective day and night in the world outside.) Thus keep in mind that circadian time in Figure 3.10 refers to subjective 'hamster time' and not objective time. In this convention, the onset of activity is defined as circadian time 12 (CT12).

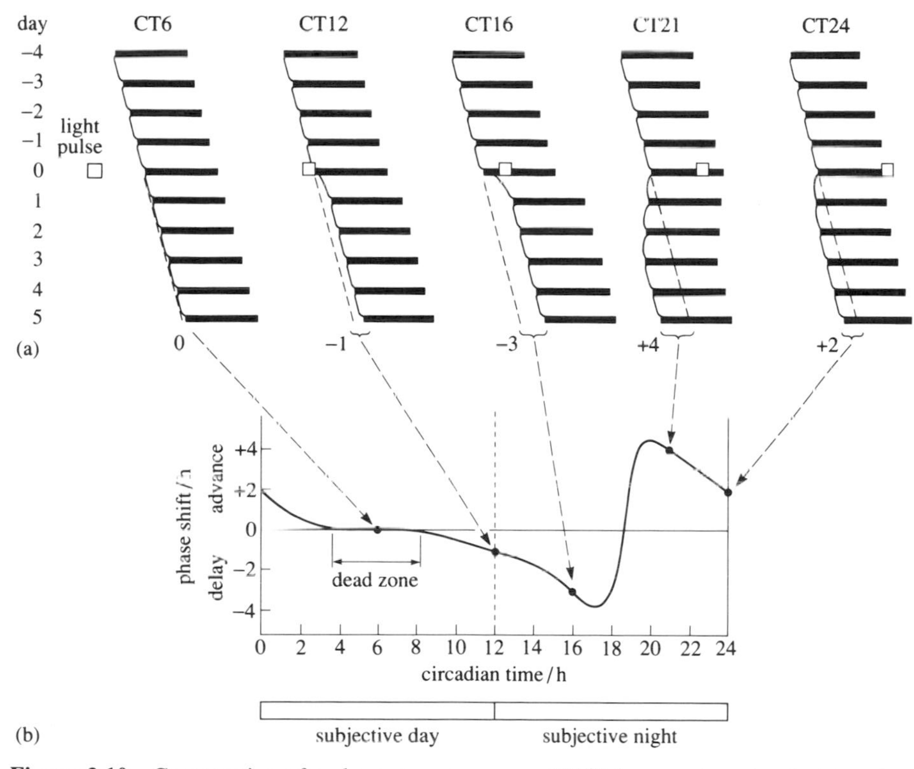

Figure 3.10 Construction of a phase response curve (PRC) for a nocturnal rodent to a 1-hour light pulse. (a) Actogram of the free-running wheel locomotor rhythm under DD. When light is given at circadian time (CT) 6, no effect is seen; this is the *dead zone*. Light given at CT12 and CT16 causes phase delays, while light at CT21 and CT24 causes phase advances. In (b) the PRC is constructed for each circadian time studied, by measuring the amount of phase shift and plotting it with respect to a 0-hour line (no phase shift, e.g. CT6). Phase advances are plotted above the 0-hour line, and phase delays below the line.

Consider first the column of results marked CT6. Note that as days progress, so the rhythm moves to the right, i.e. out of phase with the objective light/dark cycle.

- □ What does this tell you about τ?
- ■ It is longer than 24 hours.

A light pulse is given on day 0 at CT6 but it has no effect upon the rhythm. In other words, the phase shift induced by light at this time is zero and so a 0 is plotted on the graph in Figure 3.10b against CT6. Consider now an animal tested at CT12, as shown in the second column from the left. When light is given at CT12, the onset of activity is delayed by 1 hour (a phase delay) and so −1 is plotted. When light is presented at CT21, the onset of activity is advanced (a phase advance), by about 4 hours. A plot at +4 is made. As the other results show, the effect of light in inducing a phase shift is dependent on the time within the animal's activity cycle at which it is given.

Thus it can be seen that phase delays occur when the light pulse falls early in the subjective night (each side of activity onset), at a time which in nature would be around dusk. Phase advances occur when light falls late in the subjective night (at the end of activity), around the time of dawn in nature.

From looking at Figure 3.10, it can be seen how, under more natural conditions, the light/dark cycle can synchronize an activity rhythm. At about the time of activity onset (around dusk), the transition of day to night in nature, phase delays are found.

□ At the end of activity at dawn, the transition from night to day in nature, what is the phase shift?

■ Phase advances occur.

It is presumed that for species that burrow, or at least are not exposed to light during the day, the phase delays exerted by their exposure to evening light are counteracted by the phase advances of morning exposure. In this way, each day, the internal clock is reset. You might like to think of it as the activity phase being sandwiched between two forces pushing in opposite directions. If activity starts to move either way, a corrective re-entraining light exposure will restore its position. During the bulk of the subjective day, the clock is unresponsive to light and this portion of the PRC is called the *dead zone*. From a functional perspective, the clock should not be shifted around in either direction by light which the animal would naturally perceive during any excursions out of the burrow in the middle period of daytime.

3.5.2 Skeleton photoperiods

If an animal free-running in constant darkness with a period of greater than 24 hours is subjected to a single daily light pulse of between 15 and 60 minutes given at the same time of day, then the nocturnal activity period eventually coincides with the light pulse and is entrained by it. This period of entrainment does not last, however. The activity rhythm breaks away and continues to free-run again. This phenomenon of unstable entrainment is called *bouncing*.

From considering bouncing and phase response curves, it becomes apparent that not just one, but both, sensitive phases of the circadian cycle need to be stimulated by light in order to obtain stable entrainment. These two phases occur around subjective dawn and subjective dusk, and therefore it should be possible to entrain free-running rhythms using two daily light pulses applied at these times only. The complete daily light/dark cycle appears to be unnecessary to obtain entrainment;

all that is necessary is for light to fall at certain critical times in the circadian cycle. Indeed, under natural conditions, or in a laboratory where the animal has a burrow to return to, nocturnal rodents do not expose themselves to the entire daily light phase. They avoid light and only surface to 'sample the zeitgeber' for a brief period at dawn and dusk. The effect of such sampling gives rise to the notion of *skeleton photoperiods*, as modelled in the laboratory experiment described in the next paragraph. The expression means that just a skeleton of the total photoperiod consisting of two pulses is sufficient to entrain the rhythm.

Normally, the free-running rhythm of a rodent housed in DD can be entrained by two daily light pulses of only 15 minutes duration each, 6 hours apart. The phase advance caused by one pulse is exactly compensated by the phase delay exerted by the second pulse. The question that naturally arises though is this: How does the system decide which of the dark intervals between the pulses should be subjective day (the inactive/sleep time) and which interval is subjective night (the active time)? In nocturnal rodents, the longer interval between two light pulses is recognized as subjective night and time for activity. Thus, under a 13:9 cycle of two 1-hour light pulses, the 13-hour interval would be designated subjective night and the 9-hour interval subjective day.

3.5.3 The endogenous circadian clock

It is known that the brain has at least one *central* oscillator that (a) drives circadian rhythms, and (b) has a zeitgeber input. Such an oscillator is called a *pacemaker*. Depending on the species, there may be more than one circadian pacemaker in the brain, and any one pacemaker can consist of one or several oscillators. There is a circadian pacemaker in the eyes of many invertebrates (such as *Aplysia*) and also in some vertebrates (e.g. the frog, *Rana*). The so-called 'third eye', the pineal gland (see Section 3.9), is also a circadian pacemaker in some species (e.g. some birds).

There is at present much discussion as to whether the mammalian brain contains more than one circadian clock. Evidence, discussed later, shows that the central pacemaker is in the brain region termed the suprachiasmatic nuclei (SCN). However, some phenomena observed in circadian locomotor patterns defy explanation in terms of a *single* oscillator. One of the clearest of these phenomena is discussed next.

Nocturnal activity in many rodents often falls into two distinct sessions (Figure 3.11, *overleaf*). The distinctiveness of the two varies between individuals in any species; sometimes the peaks are clearly separable and sometimes they are not. The first major peak occurs early in the subjective night around dusk and has been designated as the E (evening) component. The component that occurs later, preceding dawn, is designated the M (morning) component.

This *splitting* phenomenon in mammals was first observed in the arctic ground squirrel (*Spermophilus undulatus*). These animals were being studied in the laboratory and were free-running in LL. When the light intensity in the laboratory was abruptly increased, the M component started to recur at increasingly later times, eventually forming a distinct band of activity of its own. Thus, the active component of the circadian rhythm had split into two components.

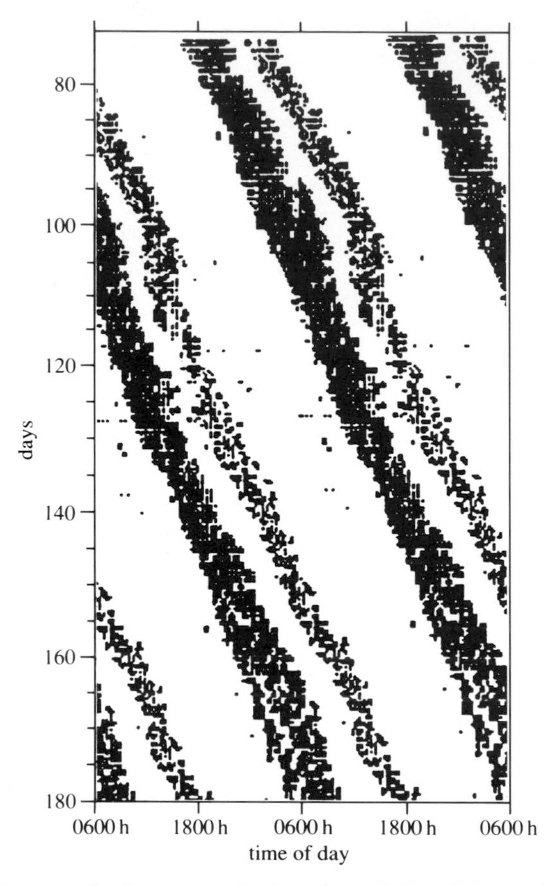

Figure 3.11 Free-running locomotor rhythm of an adult male Long–Evans hooded rat in DD. Note that the active part of the cycle actually consists of two bands of activity that persist over time.

The importance of the splitting phenomenon is that it challenges any concept of the circadian system that is built upon a pacemaker consisting of a single oscillator. The phenomenon strongly suggests that two mutually coupled oscillators are necessary to explain locomotor activity. One oscillator (M) controls the morning component of the active period and is coupled to the dark/light transition of the external zeitgeber in nature. The other oscillator (E), controls the evening component and is coupled to the light/dark transition. Joining occurs when there is a gradual return of synchronicity, and thus mutual coupling between the oscillators. These two oscillators might have periods which are different from each other. If so, under stable conditions each oscillator would exert a modifying influence on the other to produce a stable period.

Splitting is not just an artificial phenomenon found in the laboratory. It can occur naturally, as exemplified by the activity rhythm of the young trout *Salmo trutta* in the Arctic (Figure 3.12). In autumn and winter the trout is diurnal in its activity and has a single active phase. In spring, as daylight lengthens, the active phase splits into two components. One component follows sunrise (M) and the other sunset (E). By early summer, fusion occurs so that there is again a single band of activity; however, activity is now centred around midnight instead of midday as it was in winter, i.e. the trout is nocturnal in its activity. Again, these sorts of data strongly implicate two circadian oscillators driving activity rhythms and forming an important determinant of the animal's behaviour.

3.5.4 A clock for all seasons

The concept of two oscillators involved in entrainment, as previously described, allows not only for the measurement of circadian events, but also for seasonal changes, a topic that will be discussed in more detail later in this chapter. This is particularly important for seasonally-breeding animals that may have to migrate or hibernate during the winter. Seasonality has critical effects on survival and the best index for predicting seasonal change is photoperiod. Photoperiodic change over the year at any given latitude is stable and allows the organism to anticipate future climatic change. Other aspects of the climate, such as rainfall and temperature, are too unstable to be used as environmental zeitgebers on an annual basis. In the dual oscillator concept, each oscillator (E and M) records the beginning and the end of the light/dark cycle, and thus measures daylength. Each oscillator is reset independently, M by dawn light, and E by dusk light.

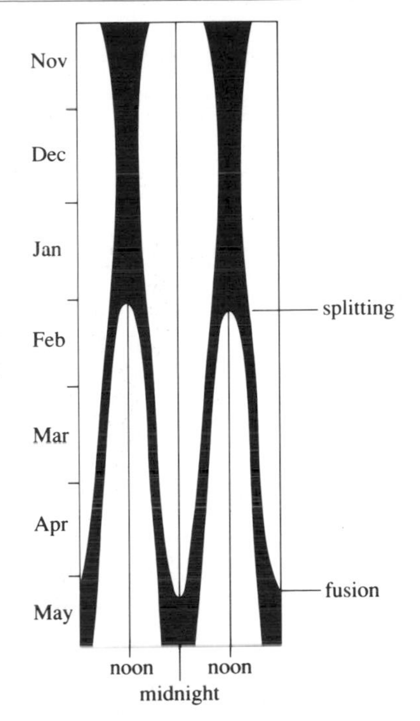

Figure 3.12 Daily activity in the young trout *Salmo trutta* in the Arctic. In October, with a long dark period, activity is restricted to daylight. As the photoperiod lengthens in spring, activity splits into two components, one following dawn, the other following dusk. By early summer, the two components fuse but are centred each side of midnight, not midday. The shaded areas indicate periods of activity.

Summary of Section 3.5

Insight into the properties of the pacemaker that drives circadian rhythms can be obtained from examining the responses of the rhythms in behaviour. This is termed a black box approach because the properties of a biological process are established without looking directly at the biological structure, but by examining its input/output performance.

Insight has been gained by keeping an animal under continuous dark and stimulating it with brief light pulses given at various points on the subjective day/night cycle. At some points in the cycle, a phase advance occurs. At other points a phase delay is seen, whereas for a large part of the subjective day there is a dead zone. From these results, a PRC can be constructed. From looking at the PRC, it can be seen how entrainment would occur naturally. Phase delays at the end of the light period will be compensated for by phase advances at the beginning of the light period. In effect, activity in a nocturnal animal will be sandwiched between these two influences.

For an animal adapted to continuous dark, two pulses of light with an interval of, for example, 6 hours between them, are sufficient to entrain activity. The longer period between the pulses is 'designated' night, and the shorter is day.

There is consistent evidence that the various circadian rhythms are driven by a single central oscillator, termed a circadian pacemaker, since it sets the pace of the body's various overt rhythms and is itself entrained by zeitgeber inputs. However, some evidence suggests that there may be two such central oscillators, which normally (i.e. when activity shows one session every 24 hours) act together, to serve as a single pacemaker. The evidence for two central oscillators consists of the observation that nocturnal activity in many rodents consists of two distinct sessions. It is suggested that each might be controlled by a different central oscillator.

The next section considers circadian rhythms in humans and their relation to the sleep/wake cycle.

3.6 The human circadian clock and the sleep/wake cycle

3.6.1 An index of the clock

Humans are not hamsters! They do not always follow the internal rhythms of their body that tell them, 'you are tired, go to sleep'. Thus, to find a good 'hand' of the human circadian clock, a more reliable measure of rhythmicity is needed. The one most used, because it is relatively easy, is core body temperature, which is usually measured continuously by means of a rectal probe. Under normal entrained conditions, the core temperature rhythm maintains a characteristic phase relationship to the sleep/wake cycle. Temperature starts to decrease several hours before the start of sleep and reaches its lowest point near the end of sleep. It then begins to rise again during wakefulness to peak late in the afternoon (Figure 3.13). This rhythm persists during sleep deprivation but the amplitude is reduced. Normally, being awake and active during the day increases the peak values while being asleep and at rest during sleep decreases the trough values.

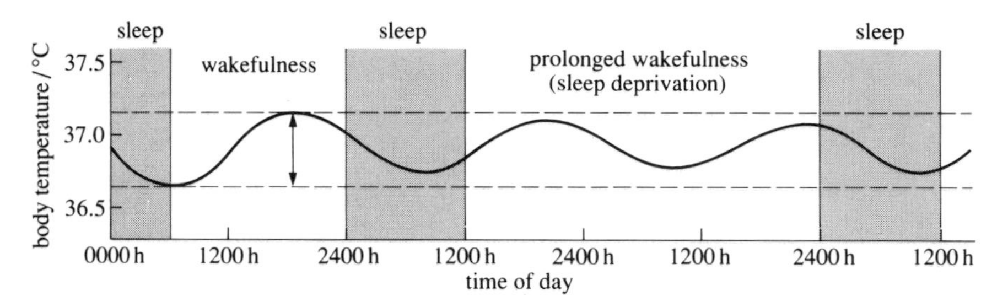

Figure 3.13 Diagrammatic representation of the circadian body temperature rhythm in humans. The arrow indicates the normal distance between the peak and the trough of the oscillation.

Another 'hand' of the circadian clock is the 'urge' to sleep, which can be represented as fatigue or alertness. Figure 3.14 demonstrates these characteristics during an experiment in which volunteers stayed awake for 3 days. Subjects were assessed every 3 hours for their level of fatigue and alertness.

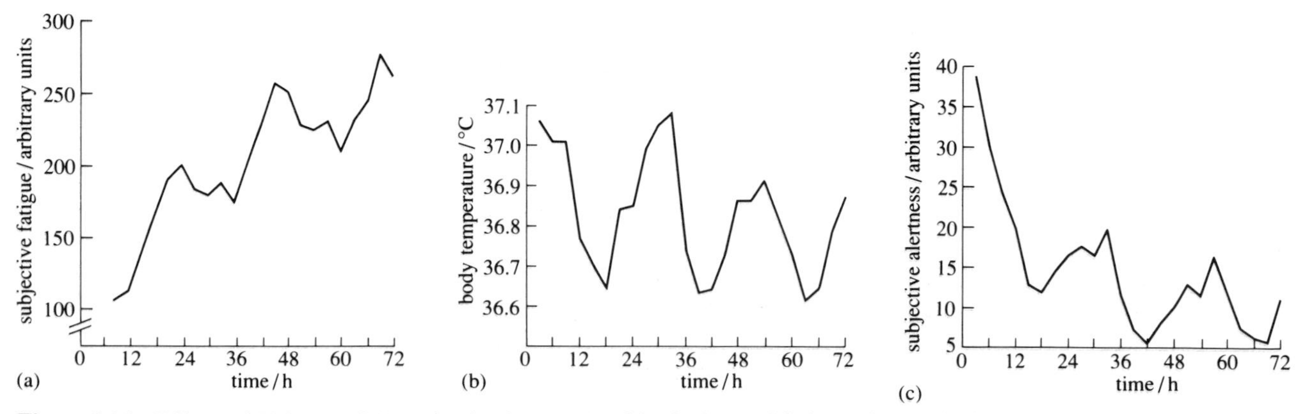

Figure 3.14 Effects of 72 hours of sleep deprivation on (a) subjectively rated fatigue, (b) core body temperature, and (c) subjective alertness. Results shown in (b) and (c) are from the same experiment involving 15 subjects. Results in (a) are from an earlier experiment involving 63 subjects.

□ Is there any evidence of a rhythm in fatigue and alertness?

■ Figure 3.14a shows that fatigue was lowest in the early afternoon and highest in the early hours of the morning, whereas alertness showed the opposite rhythm.

□ What is the relationship between body temperature and these rhythms?

■ The rhythm of alertness followed the body temperature rhythm (Figure 3.14b), so in fact the latter can be used as an index of 'sleepiness'. When temperature was at its lowest, alertness was at its lowest, and sleepiness at its highest.

□ Can the rhythm account for all of the change in fatigue/sleepiness?

■ No. The rhythm in fatigue/sleepiness is superimposed upon an increasing baseline level over the 3 days, as the sleep deprivation accumulates. This, therefore, suggests that normal sleep has two components: (1) a circadian component, and (2) a homeostatic component (Book 1, Section 7.2.2) dependent on the length of time the individual has been awake.

3.6.2 Free-running rhythms in humans

When humans are permitted to choose their times of waking, sleeping and turning lights on and off, under conditions of isolation from external time cues, the period of the sleep/wake cycle usually lengthens to an average of 24.5 hours. The internal phase relationships between rhythms also change: the low point of the temperature rhythm advances from the end of sleep to the onset of sleep. Normally, under constant conditions, different rhythms within an individual free-run with the same τ, so that a stable phase relationship between rhythms is kept. However, depending on how long the subjects live in temporal isolation, their age, as well as their psychological neuroticism, they can exhibit a fascinating phenomenon, as follows.

When German researchers began their human experiments in the underground bunker at Andechs, they found that the sleep/wake cycle could spontaneously adopt a different period from the circadian temperature rhythm. They called this phenomenon *internal desynchronization* (Figure 3.15, *overleaf*). In this particular case, the subject's sleep/wake cycle and core temperature rhythm free-ran with a τ of 25.7 hours for the first 14 days in the isolation unit. After day 14, the sleep/wake cycle suddenly lengthened its τ (and therefore slowed in frequency) to 33.4 hours, while the temperature rhythm shortened slightly to 25.1 hours. This remarkable phenomenon persisted until the end of the experiment. This meant that one rhythm 'lapped' the other more than once, so to speak, so that one rhythm had completed more circadian days than the other.

The German studies found internal desynchronization to occur in less than a quarter of their human subjects. Later studies in the United States left subjects free-running for periods longer than 2 months, and found internal desynchronization occurring much more often. They considered it possible that most humans would be susceptible to internal desynchronization provided they remained long enough in temporal isolation. This phenomenon appears to be specific to humans and has not so far been found in other animals, thus limiting its study.

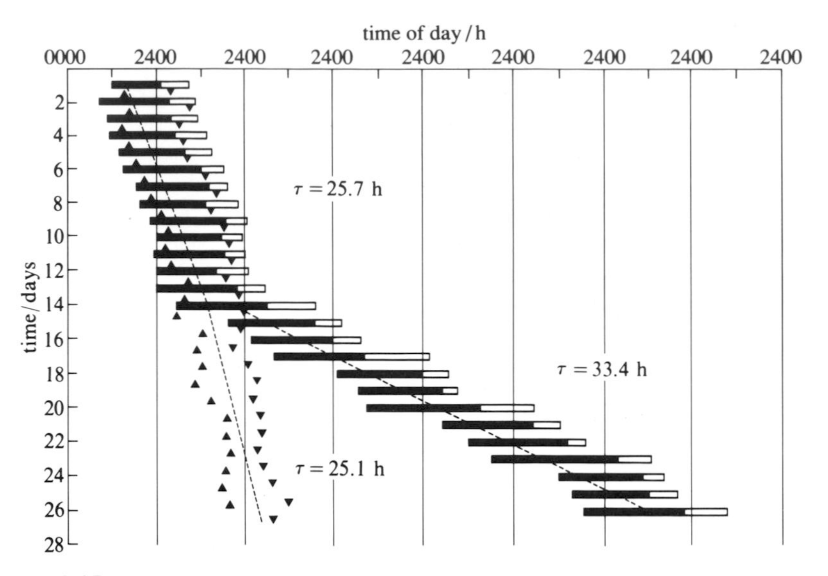

Figure 3.15 Internal desynchronization between the sleep/wake rhythm and rectal temperature in a woman in the isolation unit. Dark bars indicate wakefulness, and white bars sleep. ▼ indicates the temperature minimum, ▲ indicates the temperature maximum. Internal desynchronization occurred spontaneously on day 14.

In the desynchronized state, the two rhythms move in and out of phase with each other. Thus in fact, the free-running subjects sleep at different phases of their circadian temperature rhythm. When researchers analysed the bunker data from Andechs, and the free-running data in the American temporal isolation unit, they both found a beautiful regularity: sleep duration depended on circadian phase. Some subjects showed an extraordinary duration of 14 hours sleep and 26 hours awake. However, long sleep spans took place only when sleep onset was near the maximum of the core temperature rhythm. Short sleep occurred when sleep onset took place at or just after the low point of the temperature rhythm. This variation of sleep duration with the phase of the core temperature rhythm is illustrated in Figure 3.16. It is one of the main reasons for the difficulties in catching up on sleep that workers find after the night shift, because they cannot sleep for a long time around the minimum of their temperature rhythm.

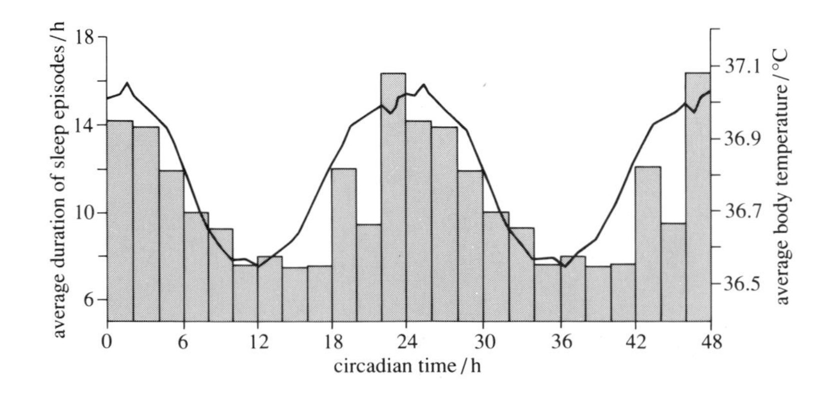

Figure 3.16 The relationship between sleep length (shaded vertical bars) and phase of the temperature rhythm. The longest sleep spans occur when sleep begins at the peak in the temperature curve while shortest sleep spans occur when sleep begins at the trough. Note tl close relationship between the descending part of the temperature curv and sleep length but the less close relationship on the ascending part.

An important finding from this study was that there was no significant relationship between sleep duration and the length of time the individual had previously been awake. This contradicts the intuitive assumptions that your length of sleep is determined by how long you have been awake. Instead, the argument is that the major determinant of the length of sleep in humans lies in the phase relationship between the circadian pacemaker and the timing of sleep. This explains, in part, the failure of sleep deprivation studies, carried out several decades ago, to demonstrate a recuperation of the sleep debt exerted by lack of sleep. Generally speaking, whether a subject was deprived of sleep for 3 days or 10 days, total sleep time on the first night of recovery rarely exceeded 11–16 hours, e.g. there was no exact depletion : repletion relationship.

The phenomenon of internal desynchronization led to the concept that there must be at least two central circadian pacemakers in humans—one driving the sleep/wake cycle and the other the temperature rhythm. However, only one such central circadian pacemaker, that in the suprachiasmatic nuclei in the hypothalamus (see Section 3.8.2), has been unequivocally identified.

3.6.3 Resetting the human circadian pacemaker

It was always assumed that the PRC in humans would be basically similar to that of all other mammalian species. It is obvious that trying to produce a human PRC in an analogous way to that in which it is obtained in nocturnal rodents would present difficulties. First, it is hard to find sufficient volunteers to stay in isolation in a bunker for long enough to receive single light pulses, every 2 weeks or so, of sufficient number to cover the entire circadian cycle. Also, most people could not tolerate continuous dark for very long! However, a number of heroic attempts have been made to obtain data for a human PRC. The shape of the curve is still a matter of debate, although the principles, phase advances to dawn light and phase delays to dusk light, remain true. Since the conditions under which human phase response curves have been determined are not exactly comparable, and since the human sleep/wake cycle is not a precise marker of the circadian clock, the complexities of human studies need to be recognized. However, excitement at the possibility of using light, a non-drug treatment, to synchronize and phase shift circadian rhythms (with the potential of treating some human sleep and mood disorders), has resulted in a remarkable expansion in basic and applied studies.

3.6.4 The sensitivity of the human circadian system to light

The early studies in the Andechs bunker used normal room lighting to manipulate circadian rhythms. Researchers found that it was the exception when a subject's core temperature rhythm was entrained to the light/dark cycle while the sleep/wake cycle free-ran. They concluded that social zeitgebers were more important than light.

□ What might constitute a social zeitgeber under normal conditions?

■ Work, friends, family, television or radio news reports, buses running, are some examples.

It is now known that these findings can be understood in relation to the light intensity (usually expressed in units called lux) used in the bunker experiments.

Humans require much higher intensity light (more than 2500 lux, equivalent to e.g. sunlight) than any other diurnal mammal to affect the circadian system. In addition, there is a variation in sensitivity to light between individuals. Thus, the few subjects who entrained their core temperature rhythm with room lighting (approximately 300 lux) were probably supersensitive to light. Later experiments using 5 000 lux intensity light indeed showed a powerful effect on the human circadian system.

This finding initiated a new era of research into the effect of light on the human circadian system. The use of dim light (less than 150 lux), sufficient for reading and other activities in the free-run situation of temporal isolation, was now ethically and emotionally acceptable as background illumination to measure the human PRC to bright light. This continuous dim light may be considered as biologically equivalent to DD in nocturnal rodents.

3.6.5 Human phase shifts with bright light

Very few technical facilities for temporal isolation studies in humans exist. Therefore, initial studies looked at the effects of multiple pulses of bright light (more than 2 500 lux) on the phase position of a marker circadian rhythm measured under entrained conditions. It was found that large phase shifts could be induced.

A group of researchers using a temporal isolation unit at the University of Manchester have looked at the effects of a single 3-hour light pulse of 9 000 lux intensity on human subjects under free-running conditions. This study demonstrated a 'classic' PRC to a single light pulse in humans that is similar in wave form to that found in other species. Also, the amplitude of the phase shifts to a single light pulse are of the same order of magnitude (1–3 hours) to that found in other species. Given that the average circadian period in humans is 24.5 hours, this size of phase advance should suffice to entrain to the 24-hour day.

One important difference exists, however, between humans and other species, with respect to the circadian phase at which light pulses are given. The human rest–activity cycle, as already noted, is not a precise marker of the circadian clock, as it is in other species. Although the principle of phase advances to dawn light does hold with respect to the human rest–activity cycle, this is not the case for the principle of phase delays to dusk light. Using the core body temperature rhythm as a marker, phase advances are elicited by light given *after* the temperature minimum around 0500 hours. However, phase delays occur only when light is given just *before* the temperature minimum, which is much later than dusk. It would appear that the timing of the PRC in humans is shifted to about 4.5 hours later than in species that use the rest–activity cycle as a marker. This definition of the timing of the sensitivity of the human circadian system to light has important consequences for strategies aimed at treating potential circadian rhythm disorders.

Summary of Section 3.6

Imposing a period of sleep deprivation on humans is associated with an increase in fatigue as a function of time, but the effect of the circadian rhythm is superimposed upon this. Therefore, to get a measure of fatigue, two pieces of information are needed: (a) length of deprivation, and (b) phase of the circadian rhythm.

One of the commonly observed circadian rhythms in humans is core body temperature. Under normal entrained conditions, the rhythm in body temperature maintains a particular phase relationship to the sleep/wake cycle. Under conditions of isolation from external zeitgebers, the temperature and sleep/wake rhythms can show internal desynchronization. Under these conditions, the duration of sleep depends upon the time of onset relative to the rhythm in body temperature. This is short if onset coincides with the low point of the rhythm in body temperature. Length of time awake prior to falling asleep is a poor predictor of length of sleep.

Under free-running conditions, humans are sensitive to light in the sense that rhythms can be shifted by pulses of light. The results and insights discussed in this section have practical applications to a number of human disorders that relate to circadian rhythms, the topic of the next section.

3.7 Circadian-rhythm-related disturbances

There are a number of disturbances in human behaviour and physiology associated with entrainment problems. These include certain forms of sleep disorder, the phenomenon known as 'jet lag', adaptation to rotating shift-work, and adaptation to permanent night-shift work. In addition, some drugs used to treat psychiatric and mood disorders have the ability to modify the circadian period of human sleep/wake and animal rest/activity rhythms. It has been suggested that part of the therapeutic mechanism is due to their effect on the circadian clock. Therefore, some researchers have suggested that alteration of biological rhythms by manipulation of the light/dark cycle could have benefits as a non-pharmacological alternative treatment in such disorders.

3.7.1 Sleep disorders

In some disorders of sleep, the sleep-generating mechanism is intact and the problem is related instead to the circadian clock; sleep is programmed at inappropriate times. This is an important diagnostic point but misdiagnosis by physicians unaware of circadian biology has probably led to underestimates of 'clock problems' among people complaining of insomnia.

Two sleep disorder syndromes that are pertinent here are the Delayed Sleep Phase Syndrome (DSPS) and the Advanced Sleep Phase Syndrome (ASPS), illustrated in Figure 3.17 (*overleaf*). Typically, in DSPS, the individual's clock is entrained much later than is socially usual (and acceptable): they cannot get to sleep till around 0300 hours and have difficulty waking up before 1100 hours. Clearly, it would be easy to misdiagnose DSPS as a sleep onset insomnia, i.e. the inability to fall asleep. In the much rarer form, ASPS, the individual's clock is entrained earlier than usual: they fall asleep in the early evening around 2000 hours and awake too early, around 0400 hours. This clock problem would be easily misdiagnosed as terminal insomnia (early morning awakening).

The distinction that is being undertaken here has important treatment implications. Neither DSPS nor ASPS will respond to conventional pharmacological or psychological intervention for sleep problems because they are clock problems, rather different from the kind of disorder usually treated under the sleep disorder heading. Neither hypnotics nor relaxation techniques will help. DSPS is quite common and is particularly taxing on the individual because, as the week progresses, a sleep debt is incurred. Societal norms make the sufferer get up in time to reach work at the normal time in the morning yet his or her clock will not allow them to go to sleep until 0300 hours in the morning. Relief is only provided at the weekends when the sufferer may sleep in. The situation is less crucial for the ASPS sufferer because they can often go to bed early if their family permits. Nevertheless, there is the psychological cost of anxiety associated with the early morning awakening, as well as a complete curtailment of evening social life.

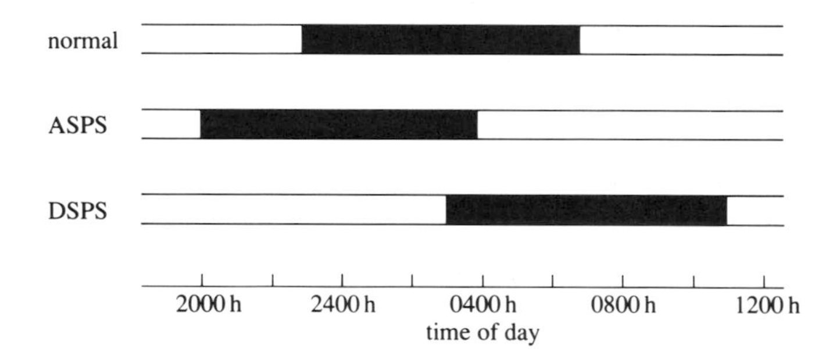

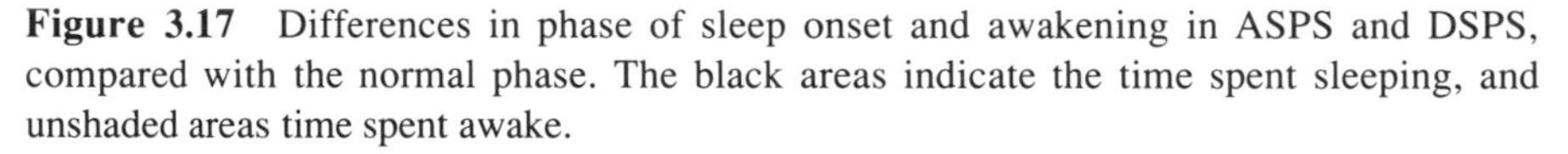

Figure 3.17 Differences in phase of sleep onset and awakening in ASPS and DSPS, compared with the normal phase. The black areas indicate the time spent sleeping, and unshaded areas time spent awake.

The cause of abnormal phases of the sleep/wake cycle in ASPS and DSPS could have two origins. First, the PRC for these particular individuals might be abnormally asymmetrical. The DSPS individuals might have an abnormally small advance area to their phase response curves or an abnormally large delay area. The opposite would be the case for ASPS. Alternatively, the PRC could be normal in shape, but the period of the DSPS clock could be abnormally long, e.g. 27 hours. Therefore, the advance portion of the PRC would be too weak to entrain properly a clock with such a long period, and there would be an inherent tendency to phase delay. In the ASPS, the period of the clock could be shorter than normal, say around 23 hours, and although the PRC would be normal, the delay portion of the PRC would be too weak to entrain such a clock. There would be an inherent tendency to phase advance. Without recourse to isolation studies of such individuals in a bunker, it cannot be determined which of the above alternatives is correct.

The DSPS sufferer needs exposure to bright morning light and avoidance of evening light by staying indoors to phase advance their clock, while the ASPS sufferer needs exposure to bright light at night and avoidance of morning light to phase delay their clock. Since natural sunlight might not be available at certain times of the year, artificial light must be used. Chronobiologists have successfully used bright artificial light, typically around 2 500 lux intensity, to treat these disorders. Remember, since indoor lighting is on average only around 200–500 lux at the most, it has very little effect on the human clock.

Having seen how to treat sleep disorders with bright light, it is a simple step to apply these same principles to jet lag and shift work.

3.7.2 Jet lag

Normally, your internal circadian system is reset every day and is entrained to a 24-hour schedule. When you are transported abruptly to a new day/night cycle, then a certain amount of time is required before internal rhythms become entrained to this new cycle (as was shown in the cockroach example of Section 3.4.2). This is what happens with intercontinental flights in the directions east–west and west–east. During re-entrainment, internal dissociation, a temporary form of internal desynchronization, occurs as different rhythms take different lengths of time to establish their normal phase relationships with each other and the new light/dark cycle. The different lags in resynchronization in different internal rhythms is subjectively experienced as 'jet lag'. At inappropriate times of the day or night one tends to feel sleepy and sluggish, alert and aroused, hungry, needing to go to the bathroom, etc. One's performance on physical and mental tasks can be impaired, and so the solving of the jet lag problem of internal temporal disarray would be of particular benefit to people travelling for competitive sport, for business, for politicians, and for airline crews in particular, and the air-travelling public in general.

From animal and some human studies, it is possible to provide rules-of-thumb which can be used to help devise schedules to speed up re-entrainment:

(1) For phase shifts in the light/dark cycle of up to 6 hours, the direction of re-entrainment should follow that of the zeitgeber, i.e. if you fly east (go 'forwards' in time), your clock will need to phase advance and if you fly west (go 'backwards' in time), it will need to phase delay.

(2) For phase shifts nearing 12 hours, the direction of re-entrainment will follow that of the individual's intrinsic circadian period. In most humans, τ is greater than 24 hours and therefore re-entrainment will be by phase delay. The reason for the delay to the 12-hour shift is that around 12 hours, the pacemaker will be at a point of indecision as to whether it should advance or delay, go east or west (the distance is identical). Therefore, the endogenous periodicity is seen as the dominant variable.

(3) The first photoperiods encountered after reaching one's destination are probably the crucial ones; once the clock has started to re-entrain in one direction, by phase delays for instance, it is unlikely that entrainment can be changed.

The events occurring during the flight have little relevance to re-entrainment as far as the air-traveller is concerned, since they are usually exposed to less than 500 lux of artificial light. However, this may not be the case for the cabin crew; the pilot may be exposed to 100 000 lux of sunlight on the flight deck.

For the passengers, the variables listed in Table 3.1 (*overleaf*) all affect or modify their comfort and the feeling of fatigue of extended travel, but they are not part of jet lag. When flying north to south or vice versa, although the variables listed in Table 3.1 apply, and there are clear *photoperiod* changes (switching of seasons across the equator), the fact that there is no light/dark cycle phase shift means there is no jet lag. The main variables governing jet lag are the size of the phase shift, the direction of the shift, and the exposure to a new light/dark cycle, particularly in the first days after arrival.

Table 3.1 Variables relevant to comfort and fatigue during extended travel.

1	Sleep deprivation
2	Enforced inactivity—boredom, fatigue
3	Meal timing
4	Alcohol ingestion
5	Seasoned vs casual traveller
6	Stress of flying
7	Health, medication, fitness, age
8	Sex differences—female menstrual cycle
9	Mood—morning vs evening types
10	Climatic changes
11	Diet changes
12	Leisure/work schedule

3.7.3 Shift-work

The chronobiological principles detailed in the last sections have some applications to rotating shift-work schedules. However, the shift-work situation is far more complex than the jet lag one in that there are a number of competing zeitgebers in the former situation. After intercontinental flight all potential zeitgebers at the destination are unidirectional, whereas for shift-work the zeitgebers are often in conflict with the demands of the shift schedule.

□ What are some conflicting zeitgebers in the shift-work situation?

■ The activity of the work place would tend to entrain the activity pattern of the worker to that particular shift. However, exposure to sunlight or noises during the daytime (when the worker is trying to sleep) would be competing zeitgebers.

Figure 3.18 shows two possible patterns of shift-work.

In the diagram on the left, the work shifts advance every week, and there are only three shifts every 24 hours. Thus, not only is the worker being asked to re-entrain in the wrong direction, but the size of the phase-shift (8 hours) is too big for the circadian clock to adapt to and the number of days given to re-entrain is too small. In the right-hand diagram, the shifts delay, they change every 2 weeks, and there are four shifts per day. In this manner the worker re-entrains in the correct direction, the size of the phase-shift is more reasonable, and ample time is given to re-entrain successfully. The last factor, amount of time given to re-entrain, could be reduced since delay phase-shifts are accommodated quite easily.

In humans, the most pertinent and obvious zeitgeber is that of social cues, e.g. the noise of people going off to work or the evening news report. Social synchronization has been shown to be effective in the German bunker. In rodents, periodic presentation of meal schedules can act as effective zeitgebers under certain circumstances.

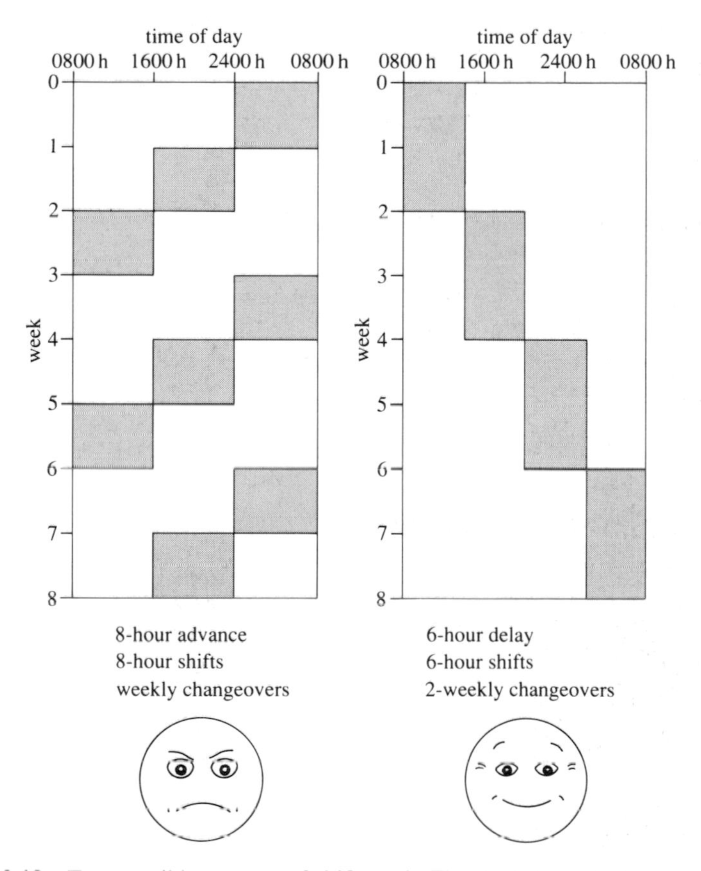

Figure 3.18 Two possible patterns of shift-work. The shaded ares indicate the time spent at work. See text for details.

Therefore, the rhythms of society and in particular the rhythms of the shiftworker's own family compete and act as 'non-photic' zeitgebers actually working against their shift schedule. Furthermore, on days off, shiftworkers often return to societal norms for socializing, waking, sleeping and eating.

This leaves two strategies for overcoming the malaise often associated with shift-work. First, since it is the night shift that is the main problem, shift rosters can be designed so that only short stays (two or three nights) are spent on night-shift. The idea here is that the individual will have returned to day-shift before the clock has started to re-entrain to nights. While this avoids the re-entrainment problem, it will result in partial or complete sleep deprivation because of the inability to obtain adequate sleep during the day. Second, circadian principles can be applied to the shift roster:

(a) Any rotating shift must be a delay shift, and not an advance one. Advance shifts are a disaster for most people (Figure 3.18).

(b) Shifts should not be rotated too frequently (except, perhaps, for the night-shift). Time has to be given for re-entrainment to take place, otherwise, even though a sensible delay schedule may be imposed, the worker is still in a permanent state of internal desynchronization; as the individual nearly re-entrains to one roster they are moved on to the next.

(c) The number of shifts over the 24 hours needs to be optimized. Having three shifts, such as morning, afternoon and night, means that 8-hour phase shifts have to take place. By increasing this to four shifts, such as morning, afternoon, evening and night, phase shifts are reduced to 6 hours and so entrainment will take place more quickly. Obviously, a further reduction to 4 hours would be even more beneficial from the chronobiological standpoint but impractical in terms of industrial economics. Finally, once a sensible shift-work roster schedule has been designed, times for avoiding or seeking exposure to natural outdoor bright light can be determined to optimize the rate of adjustment to the new schedules.

Summary of Section 3.7

A knowledge of circadian rhythms is necessary in order to understand a number of problems experienced by humans. Some sleep disorders are a consequence of an abnormal phase relationship between the light/dark cycle and the programming of sleep. A knowledge of the PRC enables a prescription of light exposure to be made. Moving in a direction east–west involves a phase shift in the light/dark cycle to which the traveller is exposed. Bodily circadian rhythms take time to re-entrain to the new light/dark cycle. Recommendations for optimal timing of shift work arise from a knowledge of re-entrainment rates and potential timing of exposure to light to augment phase shifts.

Section 3.7 completes the discussion of the properties of the circadian rhythm obtained from observation and looking at the input/output responses of the system to environmental manipulations. The remainder of the chapter is concerned with investigating the mechanism of the clock by looking at the underlying processes in the nervous system.

3.8 Generation of circadian rhythms

This section assesses what is known in mammals about the anatomical location of the clock itself, about its rhythm-generating mechanism, and some of the output pathways. How is insight gained into these brain mechanisms? One traditional method is to lesion or chemically or electrically stimulate selected parts of the brain. There are probably a number of brain regions which, when interfered with, would modify the *output* from the clock in terms of the phase, amplitude and period of the variable being measured, e.g. locomotor behaviour. However, the interference would not necessarily alter the clock mechanism itself.

3.8.1 Circadian clock location: early attempts

Over many decades, the psychobiologist Curt Richter attempted to locate the circadian clock. Working at Johns Hopkins University, Baltimore, mainly with rats, he manipulated the major hormonal systems without any major changes to running wheel locomotor rhythms. He carried out a whole series of experiments disturbing the CNS both indirectly as well as directly by lesioning. The only regions where lesions led to total disturbance of the locomotor, drinking and feeding rhythms and the appearance of arrhythmia (loss of rhythm) were the ventral and medial parts of the hypothalamus. Lesions to other parts of the brain led to interruption of the behavioural output but not to changes in periodicity.

3.8.2 The suprachiasmatic nuclei

Although the hypothalamus is a small area of the brain, it is immensely complex anatomically and is involved in a multitude of functions such as temperature regulation, reproduction, hunger, thirst, aggression, etc. Therefore, Richter's narrowing down of brain areas likely to be involved in circadian rhythm generation to the ventral and medial parts of the hypothalamus, was useful, but not specific enough.

Specificity was not attained until 1972 when researchers measured the locomotor activity and drinking rhythms of female rats whose ovaries had been removed, as well as the rhythm of the concentration of the hormone corticosterone (see Section 5.3) in the plasma, which is the liquid part of the blood. They discovered that lesioning of one specific set of paired nuclei in the hypothalamus resulted in elimination of the particular rhythms being studied. These are the suprachiasmatic nuclei (SCN, Section 3.5.3).

Figure 3.19a shows the general location of the rat SCN from the inferior surface of the brain, while Figure 3.19b shows the precise position of the SCN.

The diagram demonstrates why the discovery of the rhythm-generating function of the SCN took so long. The paired SCN are small and therefore hard to lesion precisely but, in addition, they lie just above the optic chiasma. Any lesion that is too large, or of the right size but too deep, could interfere with the visual pathways at the chiasma. Since the light/dark cycle is the dominant zeitgeber, rhythmic changes as a result of such misplaced lesions would be due to interruption of entrainment mechanisms by light and not the effect of damage to the pacemaker itself. Therefore, work on the SCN is very difficult due to the exacting requirement of placing lesions accurately.

□ Why are the suprachiasmatic nuclei so called?

■ Because they are located just above ('supra' to) the optic chiasma.

In spite of the technical difficulties, pioneer researchers in the late 1970s and early 1980s such as S.–I. Inouye, H. Kawamara, G. E. Pickard and F. W. Turek, and subsequently many others, were able to demonstrate that it was the SCN that generated circadian rhythmicity. After lesions to the SCN, the nocturnal predominance of running wheel activity and drinking was eliminated. Whereas approximately 95% of these activities took place in the dark period of the light/dark cycle prior to lesioning, subsequent to lesioning they were evenly distributed, 50% in the light and 50% in the dark. Since the time of these early experiments, there has been a growing catalogue of rhythms abolished after SCN lesions. It is important to ensure that any change to circadian rhythmicity is permanent and not a transient one due to immediate post-operative disruption of neural and glial organization caused by the lesions. As far as can be determined, SCN lesions result in a permanent loss of circadian rhythmicity. The bulk of research has been carried out on rats and hamsters.

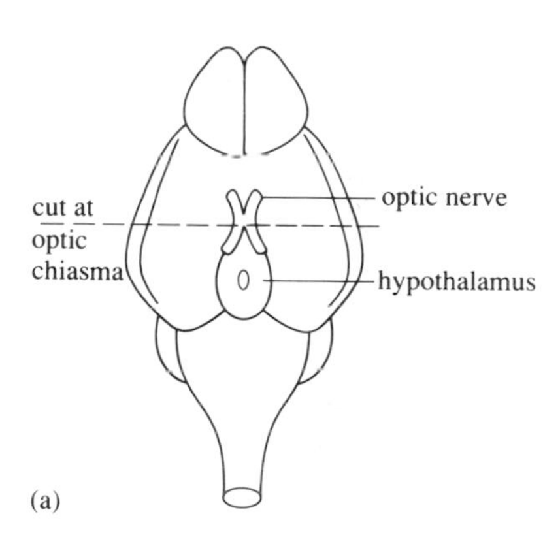

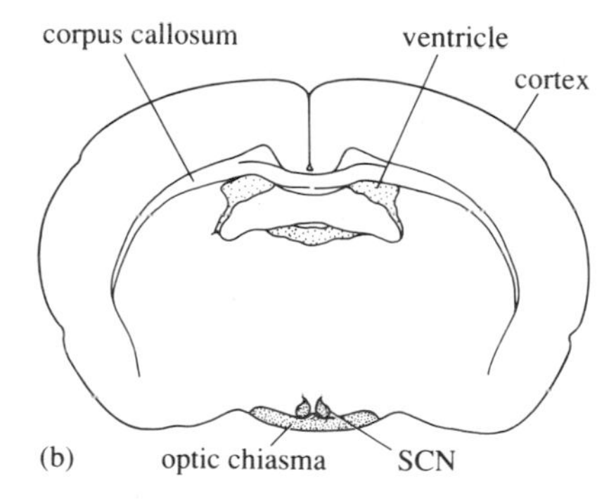

Figure 3.19 (a) Schematic drawing of the ventral surface of a rat brain, showing the position of the optic nerves, optic chiasma and the outline of the hypothalamus. (b) A section of the brain made at the point labelled 'cut' in (a), showing where the paired SCN lie just above the optic chiasma.

If the SCN are the circadian clock of rodents, the question arises as to the number of oscillators contained within the clock itself. This question has not been addressed fully, but one interesting experiment involved the elimination of splitting. This link-up with the black box experiments discussed earlier helped to substantiate the SCN as the circadian pacemaker.

Researchers have also investigated the effects of lesioning one of the SCN (unilateral lesion) or both of them (bilateral lesion) in hamsters. Figure 3.20a shows the actogram of a hamster held in continuous light for a number of days.

□ How long is the animal under these conditions before splitting occurs?

■ About 85 days.

The unilateral lesion was made at the day labelled S.

□ What are the effects of this lesion?

■ Splitting is abolished and τ is greatly reduced, to less than 24 hours.

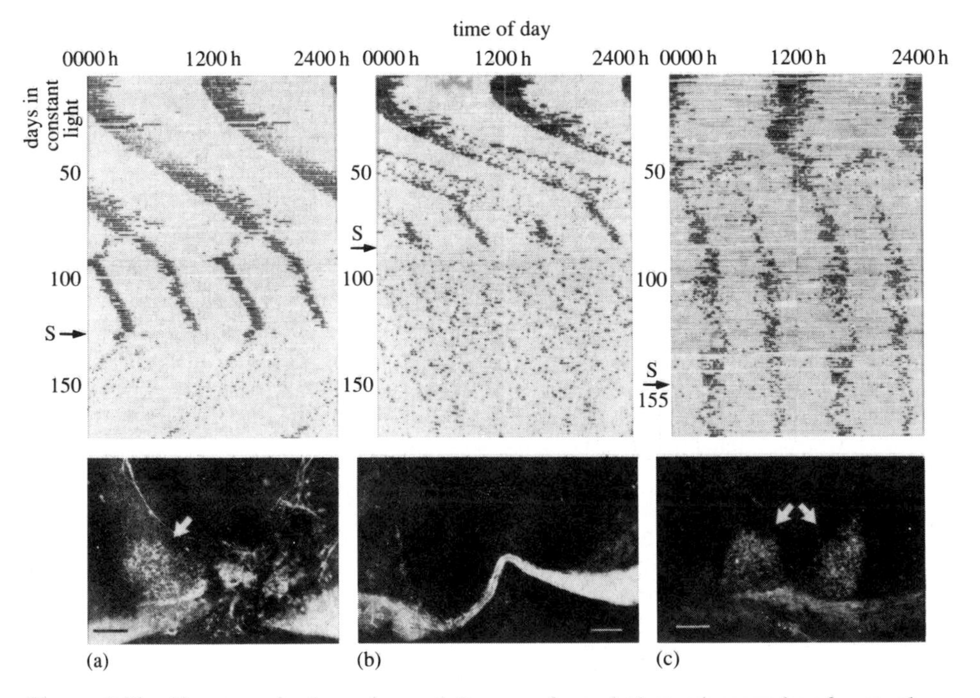

Figure 3.20 Hamster wheel-running activity records, and photomicrographs of coronal brain sections. Hamsters were kept in LL to induce split rhythms. Surgery (SCN lesioning) took place at the time indicated by the 'S' on the left of each actogram. In (a), surgery resulted in abolition of the split rhythm, with the remaining circadian band free-running with a period now less than, instead of greater than, 24 hours. The photomicrograph revealed that one nucleus was undamaged (arrow) and one was lesioned. The undamaged nucleus had retinal processes present, revealed by horeseradish peroxidase, indicating an intact retinohypothalamic tract (see Figure 3.21) input to one nucleus. In (b), surgery resulted in total disruption of running wheel activity, and correlated with the total removal of both SCN. In (c), there was no effect on the split activity rhythm of a lesion placed near to the SCN, which left both SCN (arrows) undamaged.

This indicates that the amount of SCN tissue determines circadian period, and that there must be mutual coupling between the two SCN in the non-lesioned state. In Figure 3.20b the results of bilateral SCN lesions are seen.

□ What is the effect?

■ There is a total elimination of circadian periodicity.

Figure 3.20c illustrates the specificity of the lesion: a lesion close to the SCN does not eliminate splitting. This experiment suggests that the SCN contain more than one oscillator. Presumably, the lesioning experiment does not indicate that there is one oscillator in each suprachiasmatic nucleus! Rather, the integrity of each nucleus is needed for the full generation of the activity rhythm to occur. Clearly, the SCN plays a pivotal role in rhythm generation.

The lesion studies, while highly suggestive, do not constitute conclusive evidence for the SCN being the circadian pacemaker. The SCN could be just part of an output system that couples information from oscillators in other parts of the CNS with entrainment information. However, there is other evidence, which taken together, converges to substantiate the SCN pacemaker hypothesis. The following sections consider this evidence.

3.8.3 Transplants of embryonic tissue

Research has shown that, after SCN lesions to rats and hamsters, the lost circadian drinking rhythm or locomotor rhythm could be reinstated. In the case of the drinking rhythm, a block of tissue was removed from the brains of 17-day-old rat embryos. This block, which represented most of the anterior embryonic hypothalamus containing the SCN, was implanted in the floor of the third ventricle of adult SCN-lesioned rats. Recovery of the drinking rhythm was observed within 8 weeks in the majority of animals. This type of finding clearly indicates that the neural tissue contains a system which is self-generating for circadian rhythmicity. However, even more conclusive would be the demonstration that the circadian phase and/or period of the donor rhythm is manifested in the host animal. This was done by utilizing mutant hamsters which have a very unusual short τ compared with wild-type hamsters. The period of the reinstated rhythm approximated that of the donor, rather than that of the host hamster before its SCN was lesioned.

It is not clear how the donor SCN restores rhythmicity to the host because structural integrity of the donor SCN is not needed. If an embryonic hypothalamus containing the SCN is taken, the cells separated into a suspension, and the cell suspension injected back into the brain of adult SCN-lesioned hamsters, rhythmicity is improved and even restored. However, the period of the restored rhythm is always shorter than 24 hours, whereas it is longer than 24 hours in intact hamsters. Control experiments using neural tissue from other brain regions does not improve or restore rhythmicity. Behavioural recovery after implantation was strongly correlated with the presence of certain neurons containing peptide neurotransmitter and their associated fibres, especially those containing a particular peptide called *vasoactive intestinal peptide* (VIP). Most remarkable was the finding that rhythmicity could be restored by transplants to a variety of brain sites which included the rostral midline thalamus and medial hypothalamus. Thus, the output from the implanted SCN cells may be indeed hormonal, rather than neuronal, since the correct anatomical hook-up of efferents from a normal SCN are not available from these diencephalic sites.

3.8.4 Entrainment pathways

To accept the notion that the SCN is the circadian pacemaker in mammals, it is necessary to demonstrate that the SCN has a normal sensory input from the environmental zeitgeber that entrains it. The dominant zeitgeber in mammals is the light/dark cycle, and so it would be expected that intact eyes and a central retinal projection would be prerequisites. In other words, one way of locating the circadian clock is to identify the photoreceptor and/or the entraining pathway and determine where the latter terminates in the brain. Depriving the animal of visual input, involving sectioning of the optic nerves, results in free-running rhythms; it is known that the eyes are the starting point in the entrainment pathway. It is then necessary to discover what types of retinal receptors and ganglion cells are involved in the circadian system, and to be able to follow a pathway from the retina to the SCN. A starting point is to describe the latter pathway.

Neuroanatomical studies carried out in the 1960s and 1970s concentrated on the then-known visual pathways. In rodents, removal of these pathways distal to the optic chiasma did not alter entrainment of circadian rhythms, even though the animals appeared behaviourally blind and lacked visually guided behaviour. Therefore, another pathway must be present, a pathway entirely separate from those involved in visually guided behaviour. It is possible that such a pathway is dedicated to the entrainment of rhythms.

The development of autoradiographic tracing methods (Book 3, Box 4.1) allowed the discovery of this pathway. Radiolabelled amino acids are injected and taken up into the retinal ganglion cells and are passed down the axons. Using this technique, a direct projection was found from the retina that terminates in the hypothalamus, specifically in the SCN, and was thus called the retinohypothalamic tract (RHT, Figure 3.21). The RHT projection is always bilateral (i.e. it goes to both SCNs). The RHT has been shown to exist in all mammals studied so far. (Rodent SCNs also receive a second entrainment pathway but space precludes discussion of it here.)

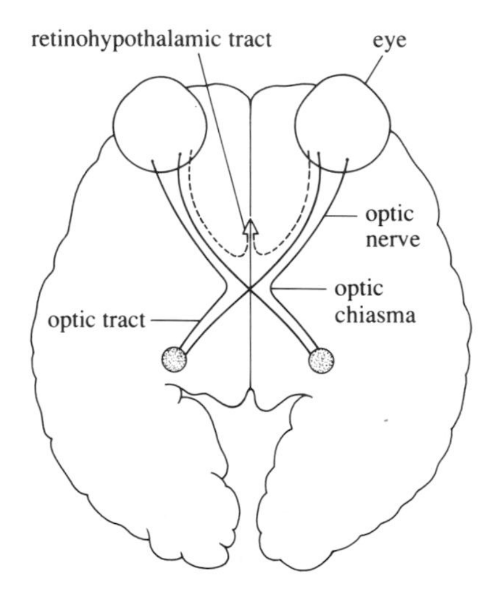

Figure 3.21 Schematic drawing of a superior view of the optic nerves and tracts to show the exit of the retinohypothalamic tract (RHT) at the optic chiasma.

Little is known of the photoreceptor mechanisms in the retina responsible for entrainment except that they are most likely to be rods and that the photoreceptor pigment appears to be rhodopsin. The circadian photoreceptor system has unusual properties, being specialized for detecting light intensity, particularly at relatively low illuminance levels compared with full daylight. It is thus ideally suited to recognize the low illuminance levels at dawn and dusk crucial for entrainment.

3.8.5 Output pathways

In electrophysiological studies, an attempt was made to see whether the SCN drove rhythms in other parts of the brain. Recordings were made from a variety of neurons in the caudate nucleus (part of the basal ganglia) and the hypothalamus to demonstrate the existence of a 24-hour rhythm in the activity of these neurons. Then, what is termed a hypothalamic *island* was made, surgically isolating the hypothalamus from the rest of the brain. The rhythm within the hypothalamic island persisted, whereas that in the caudate nucleus was lost. It would appear that the rhythm in the caudate nucleus was driven by the SCN.

Efferent projections from the SCN have been mapped using autoradiographic tracing methods. The greatest detail produced by this technique is restricted to within the hypothalamus itself. However, one of the most interesting as well as important efferent pathways involves a complicated and circuitous route from the SCN to the pineal body.

3.8.6 SCN neural activity

Recording from a number of different neurons, researchers showed that the rat SCN has a high level of neuronal activity during the day and a low level at night. A hypothalamic 'island' was made by directing a series of knife cuts rostral, caudal, lateral and dorsal to the paired SCN. Recording from neurons showed that circadian rhythmicity persisted in this preparation, where there is an absence of all neural input from the rest of the brain.

□ Does this rule out all possible inputs?

■ It rules out neural inputs. Presumably, there could still be a hormonal rhythmic input to the hypothalamic island.

This is thought to be unlikely, although not impossible.

A technique for studying the SCN totally independent of neural and hormonal inputs is to use rodent hypothalamic slices in culture. Briefly, thick coronal sections are kept functioning for a number of hours and single-cell, rather than multiple-cell, recordings are made. While at present it is not possible to follow a single neuron consistently over 24 hours, nevertheless, when day and night recordings from slices taken at different times of day are compared, a high discharge is found during the light phase and a low discharge at night. Thus, the single unit recording findings parallel the multiple unit recording findings. This important evidence supports the contention that the SCN has an inherent 24-hour rhythmicity.

3.8.7 Electrical stimulation of the SCN

Electrical stimulation of the SCN of rats and hamsters produces phase shifts of feeding and running wheel activity rhythms. If the rat SCN is electrically stimulated at different times in the 24-hour period for 20–60 seconds while under anaesthesia, and the feeding rhythm is monitored again after recovery from surgery, phase shifts are found; stimulation during the late subjective night produces phase advances while, during the subjective day, only small phase delays are found. Stimulation lateral and caudal to the SCN produces no phase shifts, nor does anaesthesia alone. In hamsters with electrodes permanently implanted in the SCN, the effect of electrical stimulation on the wheel-running activity rhythm has been monitored. Similar results to those on rat feeding were found; delays and advances occurred and these depended upon the circadian phase at which electrical stimulation took place. Both for feeding in rats and for wheel-running activity in hamsters, phase shifts induced by electrical stimulation parallel those seen in the PRC for short light pulses, suggesting that the stimulation mimics the effect of light at the neural level.

3.8.8 SCN metabolic activity

If it can be demonstrated that environmental lighting affects the SCN at a biochemical level, this would be in keeping with the requirement that a pacemaker has a direct input from the zeitgeber. Phase shifts in living animals can be shown using 2-deoxyglucose (2-DG) autoradiography (Book 3, Box 4.1).

☐ What property of the metabolism of neurons does this technique rely upon?

■ It relies on the fact that neurons in the brain are almost entirely dependent upon glucose for energy. Therefore, any area of the brain that is metabolically active will use up a lot of glucose. It follows that any administered glucose should be taken up preferentially by those brain areas that are metabolically active.

☐ If a radioactive label is attached to that glucose, what should follow?

■ Radioactivity should be concentrated in the metabolically active area.

2-DG autoradiography enables brain sections to be produced in which areas of high radioactivity, i.e. areas of high glucose metabolism, show up as dark spots.

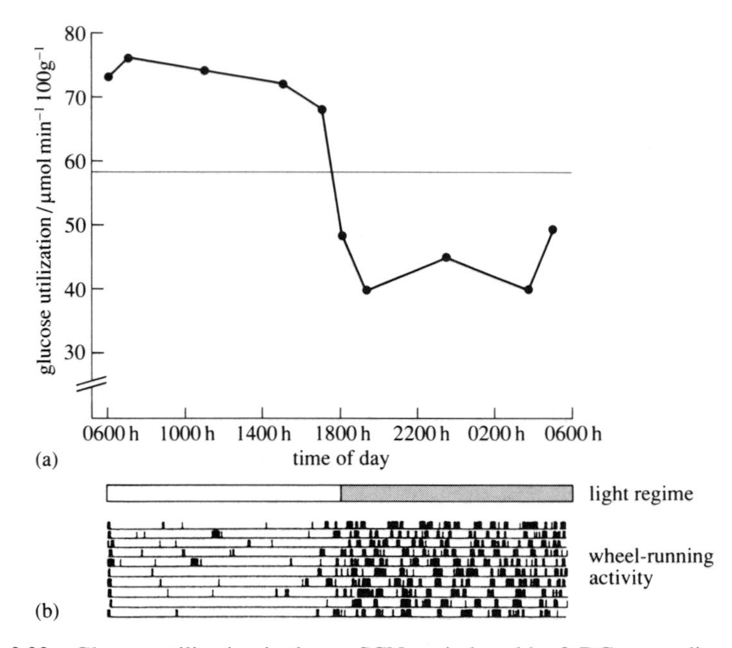

Figure 3.22 Glucose utilization in the rat SCN, as indexed by 2-DG autoradiography. (a) Glucose utilization across the 12 hours light:12 hours dark cycle. The horizontal line indicates the mean level of utilization across the whole 24-hour period. (b) Representative actogram of wheel-running activity of a single rat. Locomotor activity and 2-DG utilization are negatively correlated. In the bar representing the light regime, the shaded area indicates the dark phase, and the unshaded area the light phase.

☐ Examine Figure 3.22. What does it say about the relationship between the light/dark cycle, metabolic activity in the SCN, and locomotor activity?

- The SCN are active metabolically during the light hours of the light/dark cycle, when the animal is inactive. They are relatively inactive during the dark, when the animal is active.

However, the greatest activity was found when the laboratory light was turned on in the middle of the dark period. Thus, the SCN definitely respond to light and this response is not seen in other brain areas. This finding, therefore, is reminiscent of the phase-shifting properties of light seen in rodents (Section 3.5.1).

The rhythm in SCN metabolism is not just a reflection of environmental light/dark input. There appears to be an endogenous metabolic rhythm in the SCN. This can be demonstrated by following the development of SCN function. In the late pre-natal and early post-natal period (by day 1 after birth), rat pups develop an SCN rhythm, as demonstrated by 2-DG autoradiography. This rhythm emerges even in rats brought up in constant darkness. Most interesting is the fact that the rhythm develops before synaptic connections within the SCN develop and, therefore, the input and output connections to and from the SCN have not yet been established. It is assumed that any metabolic changes are due to 24-hour rhythmicity within the nucleus and not from stimulation coming from outside it.

Summary of Section 3.8

A number of different pieces of evidence all point to the same conclusion: that the SCN are the physical site in the brain of the pacemaker that drives the circadian rhythms of the body. Unilateral lesions of the SCN eliminate splitting, whereas bilateral lesions eliminate circadian rhythms altogether. Control lesions of neighbouring brain regions have no effects on rhythms. In rats that have lost their rhythmicity as a result of SCN lesions, rhythmicity can be restored by implants of tissue that contain the SCN. There is a direct projection from the eyes to the SCN, and this is assumed to be the major entrainment pathway. Observations of the activity of neurons in the SCN show that they have a higher activity in the light phase than in the dark. Their metabolism, as measured by 2-DG uptake, reflects this. This rhythm of metabolism appears to be endogenous. Electrical stimulation of the SCN induces phase shifts in the body's rhythms.

Although the SCN is the site of most interest in this area, the important role of the pineal gland in the phenomena of rhythmicity also needs to be considered and this forms the topic of the next section.

3.9 The pineal gland

The hormone melatonin is synthesized from serotonin (also called 5-HT) by the pineal gland. In all vertebrate species studied so far, the release of melatonin occurs mainly during the dark phase of the light/dark cycle, irrespective of whether the animal is diurnal or nocturnal in its activity. Thus, in humans and most birds, nocturnal release of melatonin occurs during sleep and rest, while in most rodents it occurs during activity. During light hours, melatonin levels are extremely low, being virtually undetectable. As can be seen in Figure 3.23, melatonin levels in human blood begin to rise late in the evening, reach their peak between 2400 to 0200 hours, and then decline in the latter half of the night so that by dawn they are back to daylight levels.

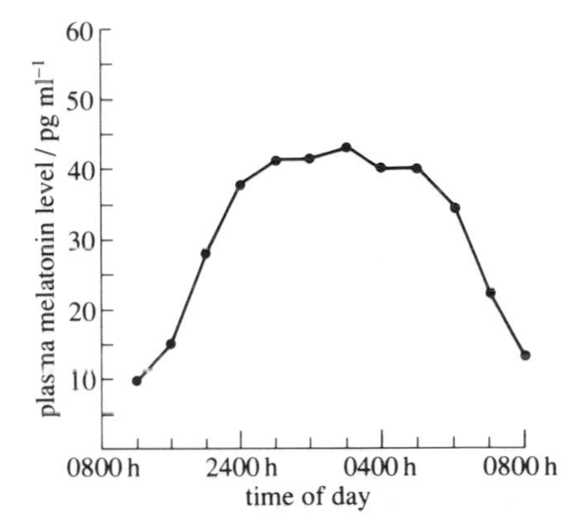

Figure 3.23 Nocturnal rise in levels of melatonin in human plasma (n = 6). Release of melatonin begins late in the evening, reaches a plateau by midnight, starts to decrease at around 0500 hours, to be back to pre-dark levels by 0800 hours. One picogram (1 pg) is a millionth of a millionth of a gram (10^{-12} g).

The melatonin rhythm is not an exogenous rhythm driven by the light/dark cycle but is an endogenous rhythm which free-runs in conditions of continuous dark. However, the rhythm can be extremely sensitive to light; a light pulse as low as 0.4 lux (equivalent to moonlight) has been shown to suppress melatonin release in certain rodents. Therefore, in continuous light, the melatonin rhythm is flattened and thus could appear to be exogenous, but this is not the case. In humans, exposure to very bright light in the evening will block the rise of melatonin.

The relationship between the pineal melatonin rhythm and the SCN is interesting. The pineal body both synthesizes and releases melatonin in the dark.

□ What is the phase relationship (Section 3.1.1) between activity of the pineal and that of the SCN (Sections 3.8.2–3.8.8)?

■ Pineal activity is 180° out of phase with that of the SCN which is most active during the light.

Loss of the SCN through lesioning results in loss of pineal melatonin production and release.

□ What does this suggest about the rhythm in pineal melatonin synthesis and release?

■ The rhythm in melatonin is driven directly by the SCN.

Since the SCN is the endogenous generator of the melatonin rhythm, this leads to a very important point: melatonin levels in plasma are a direct indicator of what the SCN is doing. In other words, although the SCN is anatomically distant from the pineal gland, we can still think of melatonin production and release as being a functional indication of SCN activity. In most species tested, melatonin release seems impervious to stress, sleep, interaction with other hormones, etc. Therefore, researchers can be fairly sure most of the time that changes in the melatonin rhythm are due to SCN changes. This gives a very powerful investigatory tool for examining the role of the SCN, particularly in humans. Melatonin level is a very precise 'hand' of the circadian clock. The question that has now to be raised is what does the pineal gland and its hormone melatonin actually do? One function has been investigated in mammals that show photoperiodism.

3.9.1 Photoperiodism in mammals

Photoperiodism in red deer was introduced in Book 1, Section 9.2.1, where the triggering of sexual activity by decreasing daylength was described. Red deer are what is termed a short-day species.

□ What do the expressions photoperiodism and short-day species mean?

■ Photoperiodism refers to biological functions that are sensitive to the ratio of light:dark in the 24-hour period. A short-day species is one in which the decreasing daylength of autumn is the trigger for reproductive activity.

In mammals, the bulk of pineal research has concentrated on photosensitive, seasonal-breeding species, particularly species of hamsters, but work on other species is increasing. For domesticated animals, an understanding of the regulation of breeding, weight gain, wool growth, etc., has significant commercial application. If seasonal changes in these variables can be mimicked, it might then be possible to market the produce to suit commercial needs in any season and not be limited to the natural seasonal rhythm.

Over the year, the golden hamster, like many other seasonal breeders, shows periods of reproductive inactivity alternating with periods when it is able to reproduce. Both sexes of golden hamster undergo marked changes to the gonads (testes and ovaries). In the laboratory, where the length of the daily photoperiod can be manipulated easily, short days are accompanied by regression of the gonads, whereas, in long days, gonadal regeneration occurs. The gonadal degeneration induced by short days is therefore not permanent; after approximately 20–25 weeks in continuous darkness in the laboratory the gonads spontaneously regenerate and become functionally mature. Pinealectomy (removal of the pineal body) prevents the short-day induced gonadal regression and, thus, it appears that the photoperiod regulates gonadal functioning, and that the photosensitive effect is mediated via the pineal gland. Under natural conditions in mid-winter, it appears that the gonads become insensitive to the pineal influence.

The annual sequence of events is shown in Figure 3.24. The decreasing photoperiod (shortening of daylength) in autumn results in hamsters spending more and more time in their underground burrows and the pineal gland induces gonadal regression. When the animal hibernates in its burrow it is in DD and light is virtually absent. The regressed gonads in winter are advantageous in that they prevent breeding and delivery of young at a time of year not ideal for survival. As spring approaches, the gonads become insensitive to the pineal's restraining influence. They start to regenerate, so that when spring arrives and hamsters emerge from hibernation, they function sexually and can breed immediately. Since their gestation period is only 21 days, the young are born at the optimal time of the year for survival. Therefore, the pineal body has acted as the principal organ for synchronizing mating, insemination, pregnancy, birth and nurturing of young.

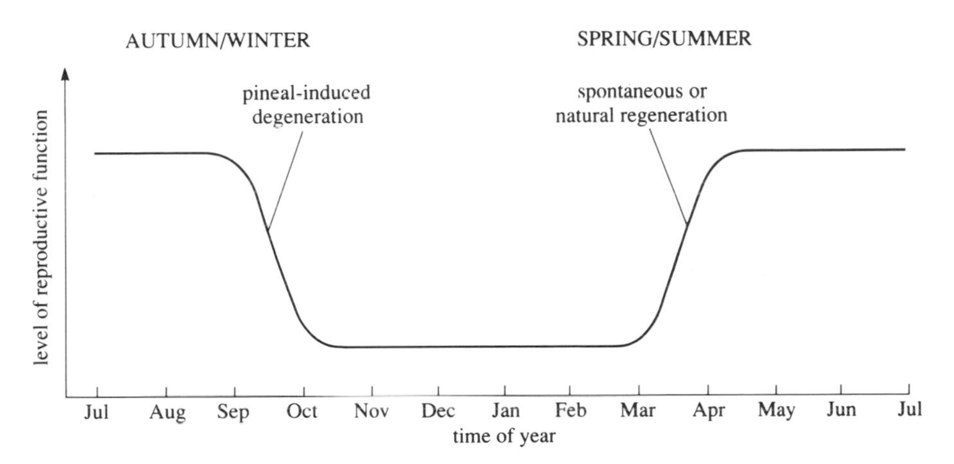

Figure 3.24 The annual reproductive cycle of golden hamsters. During short days in winter, reproductive organ growth is held in check by pineal activity via melatonin release. At about mid-winter, this pineal influence becomes ineffective and the organs begin to regenerate, so that in the long days of spring and summer, the capacity for sexual motivation and behaviour returns.

In other species where gestation length is much longer, such as the goat, the photoperiod signal for mating has to be short days, not long days (Figure 3.25); the goat must mate in autumn in order for the young to be born in spring. Therefore, from these examples it is clear that melatonin is a *timing hormone*, not, you will note, a fertility hormone.

Like circadian rhythms, such annual rhythms are anticipatory in their function (Section 1.2.3). Physiological events must be put into action before the necessary environmental changes occur. Preparation for hibernation (just as with migration) must take place before winter onset. Many aspects of a temperate climate are unreliable indicators of seasonal change from year to year, particularly rainfall and temperature changes. An 'Indian summer' and sudden harsh onset of winter would catch species unprepared if they were relying on rainfall or temperature changes as cues to the onset of winter. In nature, day length is the most stable indicator of the coming annual change in season, and many annual breeders such as those mentioned use it.

The stimulus arising from the pineal that acts as a signal to the reproductive system is melatonin. The interesting feature about melatonin when administered in the laboratory is that injection or infusion must take place in the late subjective afternoon if mimicking of the short-day effects on the reproductive organs is wanted. If given continuously in the form of an implant which releases a known amount daily, a functional pinealectomy results; the target organs cannot read the circadian signal and respond in the same way as if the pineal had been removed.

While the photosensitivity function of the pineal body in seasonal breeding mammals is both interesting and of tremendous commercial importance, the question of what the pineal does in non-seasonal breeders, like humans, remains to be answered. However, this discussion of seasonal breeding has been instructive because it has shown that, in some mammals, melatonin is a timing hormone. Thus, while in some birds and lizards melatonin is a circadian timing hormone, in some seasonal breeding mammals it is a seasonal timing hormone. Perhaps this timing function remains in humans?

Every evening at about the time you go to bed in summer, and earlier than bedtime in winter, your pineal gland starts to release melatonin into your blood stream. What does this release of melatonin do? The answer is not known but there are some interesting findings available. For instance, the pineal is not a circadian clock in mammals in the same way that it is in the sparrow and some species of lizards. Pinealectomy of rats, hamsters and other laboratory animals leads to no disruption of free-running circadian locomotor rhythms. Hence the pineal does not seem to be involved in the *generation* of circadian rhythms in mammals. However, it could be involved in synchronization of rhythms.

The outstanding feature of the melatonin signal is its reliable rhythmicity, high at night and low during the day. This chapter has already discussed how melatonin may be a good indicator of SCN activity because the rhythm is driven directly by the SCN. Therefore, using the same logic, some scientists have suggested that perhaps the melatonin rhythm is the major hormonal circadian signal from the SCN to all cells and tissues of the organism. In other words, melatonin might serve as a kind of *internal zeitgeber* to relate rhythmic activity in various regions of the body to a kind of 'master clock' in the SCN.

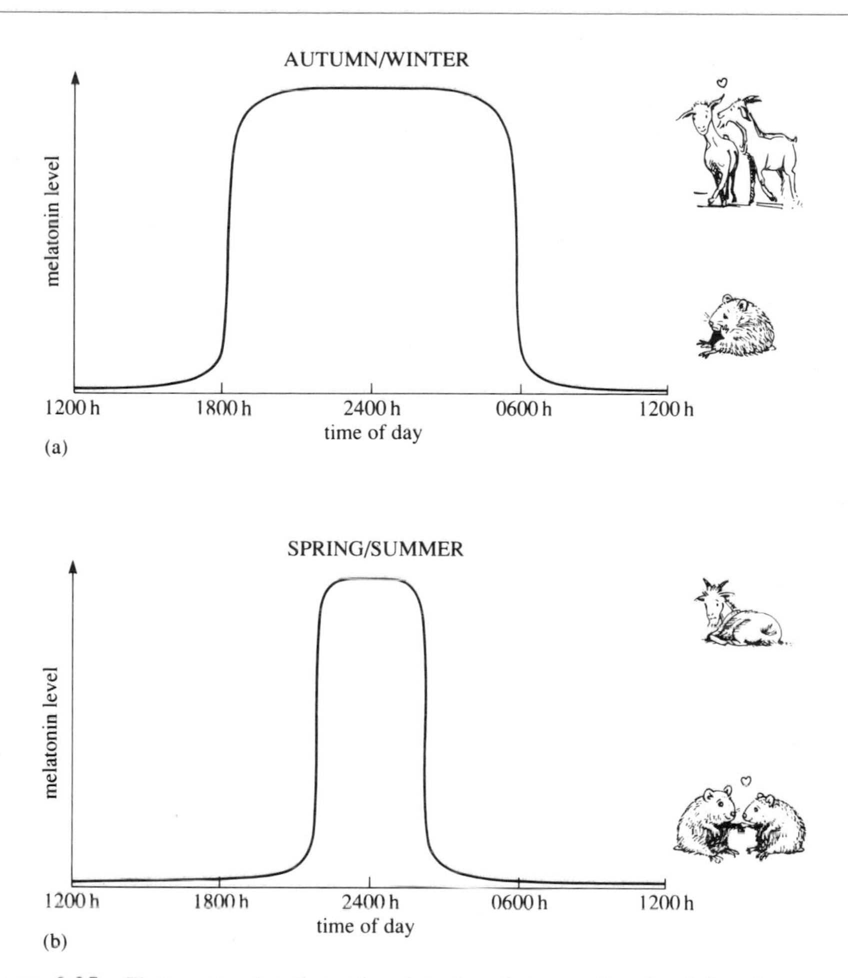

Figure 3.25 Changes to duration of melatonin release as the signal for changes to reproductive function. In autumn/winter, under short days and long nights, duration of melatonin release is at its greatest. Mammals with long gestation periods mate at this time so that the young are born in spring. In spring/summer under long days and short nights the duration of melatonin release is at its shortest. Animals with short gestation periods mate and the young are born soon after.

□ What is the significance of placing the word *internal* before zeitgeber in the description?

■ The term 'zeitgeber' usually refers to external phenomena such as the lighting cycle. Here it draws a distinction with this normal usage.

However, the only evidence available to support the speculation of such a function lies in findings of altered phase relationships after pinealectomy in a small number of CNS and peripheral tissue functions, although the pharmacological findings presented in Section 3.10.3 are also suggestive. All that can be said at present is that there exists a growing catalogue of variables showing subtle and potentially important changes found after pinealectomy, e.g. in the immune system, blood pressure, and modification of the ageing process. However, none of these changes seem to be crucial to normal functioning.

Clearly, human pineal research is at the forefront of an exciting era, with so much now known about pineal anatomy, biochemistry and physiology, and so little known of its function. The discovery of melatonin's function(s) in humans may represent the last great frontier in functional hormone research, since most other endocrine glands and hormonal systems have been well researched.

Summary of Section 3.9

The major hormone produced by the pineal gland is melatonin, which is released during the dark phase of the light/dark cycle. By looking at melatonin levels in the plasma, a good index of SCN activity can be gained. Melatonin serves a timing role in reproductive behaviour in seasonally breeding species.

3.10 The biochemistry of the biological clock

This section introduces a number of individual drugs and neurotransmitters. Do not worry about remembering their names but concentrate upon the principles of what is going on.

In Section 3.5 the way in which light exposure could be used to reset the human clock was discussed. However, natural outdoor lighting might not always be present in sufficient intensities in winter in temperate and polar zones, and it may not always be convenient to use bright artificial light even if it is available. Therefore, the development of drugs that can reset the clock would be tremendously useful as an alternative, or an adjunct, to light exposure. First, it is necessary to consider which neurotransmitters, putative neurotransmitters (meaning substances that are suggested as possible neurotransmitters), and peptides are present in the SCN and its afferent pathways. What is known of their function with respect to circadian time-keeping and transmission of input from the eyes? These questions can guide a search for drugs that will reset the clock. Conversely, the use of ever more specific drugs permits dissection of the neurochemical mechanisms underlying entrainment and rhythm generation.

A number of neurotransmitters, putative neurotransmitters, and peptides are found in the SCN, including acetylcholine, noradrenalin and dopamine. These give an indication as to possible agents (agonists and antagonists) that might be used to shift the phase of the clock, either directly by affecting neurons within the nucleus itself or indirectly by pharmacologically stimulating the entrainment pathway terminals. However, an investigator is faced immediately with two problems.

First, most of the peptides and neurotransmitters found in the SCN are not unique to this location but are reasonably ubiquitous in their CNS distribution. They are involved in the control of a number of behaviour patterns as well as physiological and endocrinological processes. Therefore, any pill prescribed to alter the clock (a *clock pill*), as in, for example, jet lag, would have to stimulate the SCN selectively, leaving the same transmitter and peptide systems in the rest of the brain untouched. In addition, the pill would have to be administered orally but survive gastric digestion, but at the same time not stimulate the peripheral nervous system and organs. Second, when light induces phase shifts, several neurotransmitter systems may act *synergistically* (meaning the effect of an individual neurotransmitter is reinforced by activity of other neurotransmitters), and this synergistic interaction could be quite complex. Therefore, administration of a single neurotransmitter or peptide may have little effect on SCN output.

Despite these barriers, there are compounds which are known to affect circadian rhythmicity, three of which are described below. Since the commercial application of any such pill is large, there will undoubtedly be an intensification of research for other compounds in the foreseeable future.

3.10.1 Neurotransmitters mediating entrainment

Early studies suggested that *acetylcholine* was involved in the entrainment pathway to the SCN, even though neither acetylcholine nor its synthesizing enzyme are found in the optic nerves or optic tracts. Acetylcholine is broken down rapidly, so a synthetic analogue called carbachol is often used to mimic its action. Injection of carbachol into the brain's ventricles in mice elicited a phase-response curve similar to that after short light-pulses. However, the ubiquitousness of acetylcholine in the CNS makes it unlikely that an anti-jet lag pill will be invented based on cholinergic principles.

Electrophysiological and behavioural studies indicate that the primary neurotransmitter of the retinohypothalamic tract is likely to be an amino acid exerting an excitatory effect. A demonstration of the physiological role of such amino acids comes from recent work in hamsters, showing that the phase-shifting effects of light pulses can be blocked by prior treatment with a drug blocking one of the glutamate receptors.

3.10.2 Rhythm generation in the SCN

There are three substances which show their highest concentration in any part of the CNS within the SCN: the neurotransmitters serotonin and GABA, and the neuropeptide arginine vasopressin (AVP). However, experiments that have eliminated almost all CNS serotonin did not eliminate behavioural circadian rhythmicity. A genetic strain of rats devoid of any AVP, the Brattleboro rat, also shows circadian rhythmicity, and replacement AVP does not alter the period or phase of the free-running period. Thus, neither of these appear to be intrinsic to the circadian time-keeping mechanism.

The neurotransmitter GABA, however, appears to play an important role in SCN function. Drugs that mimic GABA action are valproate and the benzodiazepines. Valproate affects circadian period in a number of species. Experiments showed that injections of benzodiazepines (drugs that increase the effect of GABA) facilitated re-entrainment of rat circadian rhythms after phase-shifts of the light/dark zeitgeber. In the mid-1980s researchers showed that injections of a benzodiazepine (diazepam) could block light-pulse induced phase advances of hamster locomotor rhythms, but had no effect on phase delays. In contrast, bicuculline, a drug that selectively blocks GABA, also blocked phase delays induced by a light pulse but not phase advances. These opposite effects of a drug that potentiates GABA and one that blocks GABA were taken as evidence that the entrainment input to the circadian clock was mediated or modulated, at least in part, by GABA neurotransmission.

Even more exciting were findings on hamsters involving a short-acting benzodiazepine, triazolam. Triazolam injections induced phase shifts in free-running locomotor rhythms, and these depended upon the time of injection. The PRC to triazolam is shown in Figure 3.26 and the application of these data to speeding up re-entrainment under light/dark conditions is shown in Figure 3.27.

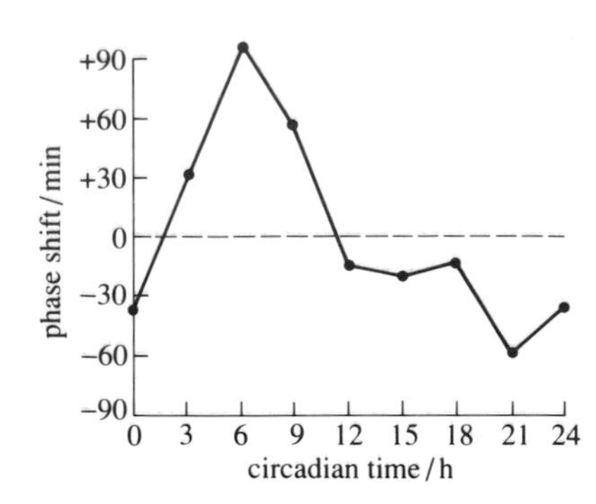

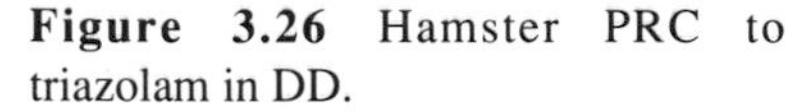
Figure 3.26 Hamster PRC to triazolam in DD.

☐ Compare the effect of light pulses (Figure 3.10) and triazolam (Figure 3.26) in inducing phase shifts.

■ Figure 3.10 shows a dead-zone during which light pulses do not induce a phase shift, whereas there is no dead-zone in Figure 3.26. Also there are times (e.g. CT21) where triazolam induces phase delays but light induces phase advances. There are other times (e.g. CT6) where triazolam induces phase advances but there is a dead zone to light stimulation.

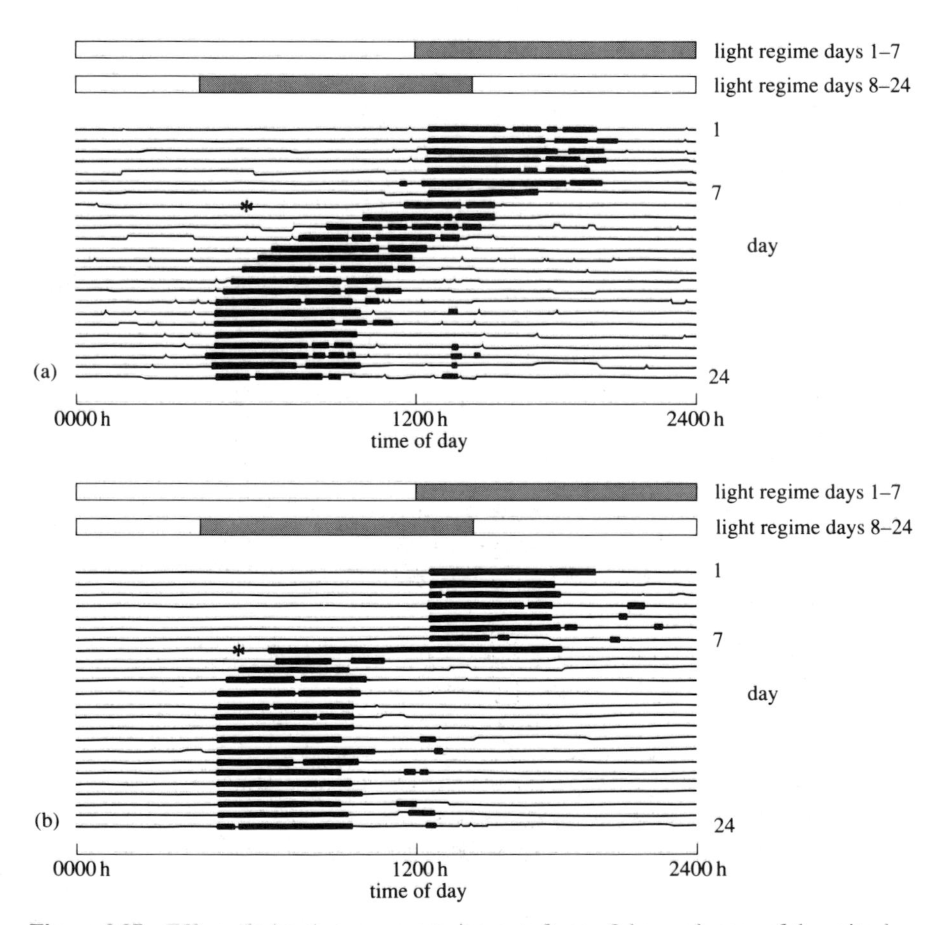

Figure 3.27 Effect of triazolam on re-entrainment after an 8-hour advance of the zeitgeber (in the bars representing the light regimes, shaded areas indicate dark phases, and unshaded areas light phases). One day after the lighting change, hamsters were injected either with a neutral (control) solution (a) or triazolam (b) at the time indicated by the star. The triazolam-injected animal re-entrained within about 3 days, while the control-treated animal took 8 days.

These differences between the effects of triazolam and light suggests that triazolam does not act on the pathway of input from light to the biological clock. The fact that the PRC to triazolam is similar under both continuous darkness and continuous light conditions also supports this interpretation of independence from light input, as does the finding that blind animals with a degenerated retinohypothalamic tract show the same PRC to triazolam as intact animals. Thus, it appears that triazolam works directly on the circadian pacemaker.

The phase-shifting findings on triazolam aroused enthusiasm but it was rather short-lived. In humans triazolam failed to improve sleep after phase shifts of the sleep/wake schedule, and one trial by the USA Airborne Corps showed impaired learning as well as failure to improve sleep after an 8-hour flight across time zones. Furthermore, there were reports that triazolam had no phase-shifting properties in rats, which again points out the importance of studies across species to gain generally valid information. The hamster is not a rat (in this regard, a rat is more similar to a human): hamsters metabolize triazolam quite differently and the behavioural effect is the opposite. In rats (as in humans), triazolam is a sedative, but in hamsters it is an activating drug. Indeed, it appears that the actual mechanism of phase-shifting by triazolam is mediated by activity. If hamsters are prevented access to their running wheels, triazolam does not induce a phase shift. The effect is dependent upon hamsters being allowed to express their hyperactivity after injection; it is probably this hyperactivity in itself that induces the phase shift.

3.10.3 Melatonin

Section 3.9 described the production and release of melatonin by the pineal gland. Where are the receptors for melatonin? Melatonin receptors in many species are highly concentrated in the SCN. Thus this appears to be an important feedback location (target) for melatonin released during the night time. What is known, then, of the action of melatonin on the circadian system?

Daily injections of melatonin to rats free-running in DD can entrain their wheel-running and drinking rhythms, provided the injections are given within a specific part of the cycle.

□ Examine Figure 3.28 (*overleaf*), which shows the effects of melatonin injections on activity rhythms. When does entrainment occur?

■ Entrainment takes place if the onset of activity coincides with the time of day that the melatonin injection is administered. After about 15 days, when the onset of activity coincided with the time of the injection, entrainment took place and was sustained until injections ceased in stage 3. Injections at times removed from this part of the free-running rhythm are ineffective.

□ Can researchers be sure that it was specifically the melatonin that induced entrainment and not, say, an activating effect of the injection procedure itself?

■ Examination of the control-injected animals shows no evidence of entrainment.

Thus, there is a very narrow window or 'gate' of sensitivity to exogenous melatonin; it is as if the circadian gate in the pacemaker is only open once every 24 hours for approximately 3 hours. This is seen more clearly after single injections of melatonin are given to produce a PRC; melatonin produces phase advances between CT9 and CT12 and no phase delays at any other circadian time.

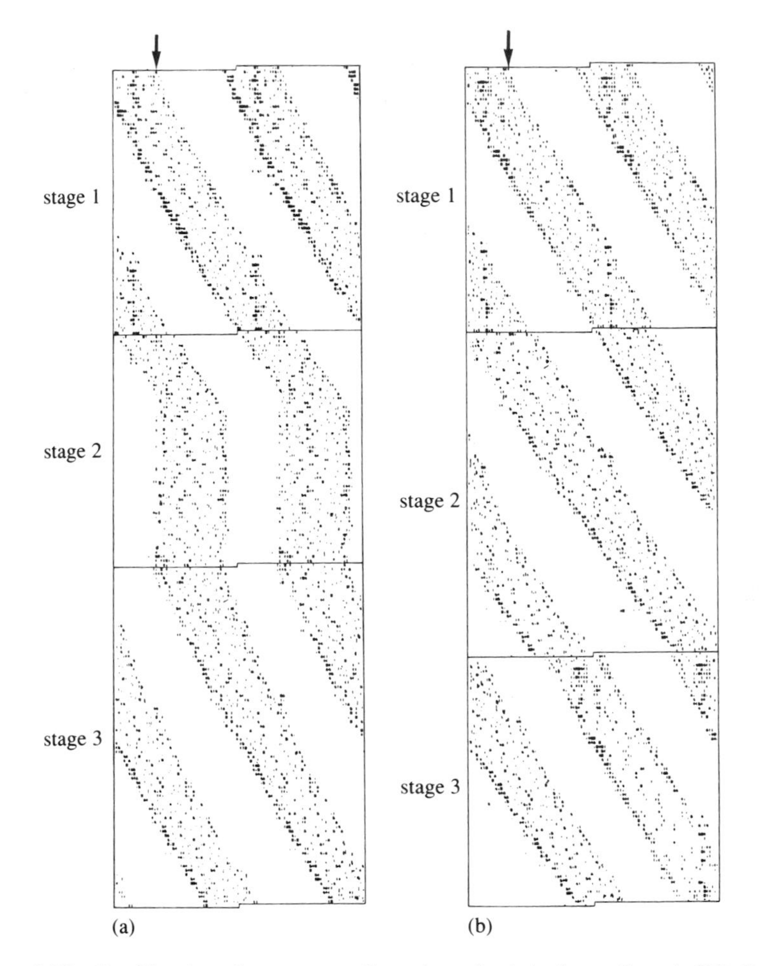

Figure 3.28 Double-plotted actograms of running wheel rhythms of rats in DD. In stage 1, rats free-ran for 60 days pre-injection. In stage 2, rats were injected daily with melatonin (a) or a control substance (b) at the time of day shown by the arrow. In stage 3, the injections were stopped, and the rats were allowed to free-run again.

On this basis, one might consider that melatonin would be useful to entrain human rhythms that are unstable under light/dark cycles, or free-running rhythms where the zeitgeber is absent or too weak. The latter situations occur in some blind people, astronauts when in outer space, and submariners who spend months underwater. The rat data would also indicate that melatonin could be useful for phase advances (as in eastward flights) but not phase delays (westward flights). Therefore, while triazolam from hamster studies looked useful as a phase-shifting agent but not an entraining agent (due to pharmacological tolerance), melatonin from rat studies seemed useful as an entraining agent but not as a phase-shifting agent (because of its unidirectionality).

Would melatonin be useful in combating jet lag for eastward flights only or for both east and westward flights? More recent evidence indicates that melatonin reduces jet lag after flights in both directions.

Summary of Section 3.10

An identification of transmitters found in the SCN would give possible pointers to drugs that might be used to change the phase of the body's circadian clock. However, any pill based upon exerting agonist or antagonist action on such transmitters meets the enormous problem that a given transmitter is unlikely to be located just in the SCN. The same transmitters might well have multiple functions involving many brain regions. Insight has been gained by comparing phase-shifting of the rhythm caused by injecting chemicals and by exposure to light. If the effect of the chemical injected mimics light exposure, this suggests it might be acting on the entrainment pathway from the eyes to the SCN.

Summary of Chapter 3

Chapter 3 reiterates the important message of Chapters 1 and 2—that to understand the control of behaviour, it is necessary to consider factors both external and internal to the animal. Furthermore, the material discussed in Chapter 3 is a good example of the way in which it is necessary to consider the subtle *interactions* between these two sets of factors. Thus the evidence shows clearly that the timing process underlying the circadian rhythms of behaviour is endogenous; the rhythms free-run when the external zeitgebers are eliminated. However, these rhythms normally entrain to external zeitgebers. Only by being clear about the nature of this interaction between external and internal factors can the results be meaningfully interpreted.

This chapter has demonstrated how insight into biological clocks has been gained by both a black box approach, and the approach of looking at, and intervening in, the nervous system. As well as giving some new information, the chapter should also have served to reiterate some familiar messages. As in Chapter 2, the problems associated with interpreting evidence on the role of transmitters was emphasized. Chapter 3 demonstrated that the use of lesions and injections of chemicals is fraught with difficulties, as will be discussed again in the next chapter.

Again with biological clocks, as with Chapter 2, a multidisciplinary perspective is needed in order to understand fully what is going on, and this involves asking questions on the causal mechanisms underlying the clock, and the function served by the clock. The topic of the function of biological clocks has already been introduced (Chapter 1), and Chapter 3 has reinforced and extended this message. The anticipatory role of rhythms, introduced in Chapter 1, was discussed again in Chapter 3 in the context of photoperiodicity.

The chapter has introduced a dynamic area of research that has made rapid progress in recent years. At the present time, the role of biological clock disturbances in such areas as depression is a topic of vigorous research.

Objectives for Chapter 3

When you have completed this chapter, you should be able to:

3.1 Define and use, or recognize definitions and applications of each of the terms printed in **bold** in the text.

3.2 Distinguish between endogenous and exogenous rhythms, and in this context, explain what is meant by 'entrainment' and 'zeitgeber'. (*Questions 3.1 and 3.2*)

3.3 Give examples of rhythms other than circadian. (*Question 3.3*)

3.4 Explain how the existence of rhythms running out of phase between individual animals can produce confusion in interpreting data. (*Question 3.2*)

3.5 Explain how a phase-response curve is constructed and relate this to natural entrainment of rhythms to the light/dark cycle. (*Question 3.4*)

3.6 Describe the problems involved in using lesion and chemical injection studies to investigate the neural basis of biological clocks. (*Question 3.5*)

3.7 Describe the evidence that points to the suprachiasmatic nuclei as being the site for the central clock, the pacemaker. (*Questions 3.5 and 3.6*)

3.8 Describe the evidence the points to the pineal body as having a role in biological rhythms.

Questions for Chapter 3

Question 3.1 (*Objective 3.2*)
In Figure 3.3, approximately how far had the rhythm drifted out of phase with the zeitgeber by day 20?

Question 3.2 (*Objectives 3.2 and 3.4*)
The following is part of an imaginary conversation between two SD206 students. Where are they going wrong?

Jane I have been observing these rats for days and they are perfectly entrained on the light/dark cycle. They sleep in the light and are active in the dark. Whenever I change the timing of the light/dark cycle, the rats re-entrain within a few days. That suggests to me that the rhythm is exogenous rather than endogenous, i.e. there is a zeitgeber in the external world.

Sean I also obtained some evidence pointing the same way. I put my rats in continuous dark. I then went in and looked at their behaviour a few weeks later. Every 3 hours for 24 hours, I looked at them, observing them for 30 minutes each time for how much they ate, drank and ran in their wheels. The mean values showed absolutely no difference between them. However, when I put the light/dark cycle back on, I saw a clear rhythm again. Thus, they are clearly driven from outside.

Question 3.3 (*Objective 3.3*)
Which of the following rhythms are evident in the results shown in Figure 3.5? (a) Circadian, (b) ultradian, and (c) infradian.

Question 3.4 (*Objective 3.5*)
With reference to Figure 3.10, at what time would a pulse of light need to be given to induce a phase advance of +2 hours? What is the frame of reference for expressing your answer?

Question 3.5 (*Objectives 3.6 and 3.7*)
Examine the record shown in Figure 3.11. How would you expect this record to be affected by (a) unilateral, and (b) bilateral lesions of the suprachiasmatic nucleus (SCN)?

Question 3.6 (*Objective 3.7*)
Examine Figure 3.22. This result supports the idea that the pacemaker is located in the SCN. Using this experimental set-up, what additional result would lend support to the hypothesis that the rhythm is endogenous rather than exogenous?

Further reading

Pittendrigh, C. and Daan, S. (1976) A functional analysis of circadian pacemakers in nocturnal rodents. V. Pacemaker structure: a clock for all seasons, *Journal of Comparative Physiology*, **106**, pp. 333–355.

Waterhouse, J. M., Minors, D. S. and Waterhouse, M. E. (1990) *Your Body Clock*, Oxford University Press.

CHAPTER 4
AGGRESSION

4.1 Introduction

The purpose of this chapter is to take a particular topic, aggression, and then explore the ways in which aggression is controlled. As with other topics in this book, the control of behaviour by external and internal factors will be examined. The questions of interest to be considered are: Why is one animal aggressive while another is not? Why is an animal aggressive in some circumstances and not others? Why are some species aggressive whilst other species are not?

Aggression is a vast topic which can be studied using a number of different approaches, two of which form the subject matter of this chapter. The first way of looking at aggression owes much to the kind of approach developed in Chapter 7 in Book 1 and Chapter 2 of the present book. This approach consists of looking at the internal and external factors underlying behaviour, and the nature of the processes that relate these factors to behaviour. Therefore, the first part of the chapter deals with aggression from a causal perspective. The second approach relies on some of the ideas about how natural selection has shaped behaviour, developed in Chapter 4 in Book 1, and deals with aggression from a functional perspective.

First, it is necessary to consider two other issues: (1) what does the term aggression mean, and (2) why are animals aggressive?

4.1.1 Defining aggression

The word aggression needs no introduction: it is familiar to us all. However, in everyday language the word is used in a variety of different contexts, and it conveys different meanings. For instance, it is used to refer to motivation, as in 'pent-up aggression'; it is used as a substitute for determination, as in 'showing a lot of aggression'; it is used to mean a specific behaviour as in 'an act of aggression'. These same three meanings are conveyed by the phrases an 'aggressive animal', an 'aggressive salesman', and an 'aggressive act'. In this chapter, the word is used in the context of competition.

In Section 4.3.2 in Book 1 you were introduced to the concept of competition and to its two forms, direct competition and indirect competition. Direct competition is where two animals compete directly with each other for access to a resource. The context in which aggression is used in this chapter is that aggression is a component of direct competition. Even within this context, it is not possible to define aggression in such a way so as to be appropriate for all animals and under all circumstances. Three principal styles of definition are:

1 By form, i.e. by the movements used, what an animal is doing, e.g. biting, goring, butting and hitting, etc. Violence, much of human aggression, and aggression studied in the laboratory are very often defined by form.

2 By consequences, i.e. by the outcome of a particular interaction in terms of a resource. If, as the consequence of an interaction, one animal gets a particular resource at the expense of another, then according to this type of definition, the behaviour shown during the interaction is aggression. For example, one sand wasp getting the nest hole is a possible consequence of two female sand wasps fighting over the same nest hole. Sole possession of part of a wood is a possible consequence of a male great tit (*Parus major*) singing to keep conspecifics off its territory. Both are examples of aggression using this style of definition.

3 By a combination of form and circumstances: particular behaviour patterns performed under particular circumstances. Behaviour shown in defence of a nest, in response to a predator, or in response to an irritating or painful stimulus are all examples of aggression using this style of definition.

In each of the styles of definition above, aggression is either a particular behaviour, or a collection of behaviour patterns; the terms aggression and aggressive behaviour are therefore synonymous and will be used interchangeably. For now, aggression will be defined by form to mean fighting—the vigorous, physical contact between two or more individuals where the participants use whatever weapons they have against their opponents.

There are two further points to note. In the discussions that follow, only interactions between conspecifics are considered; this restriction both focuses the discussion and excludes predation, which is better thought of as a component of feeding behaviour, at least from the predator's point of view. The second point is that you should not worry that a rigorous, all-embracing definition of aggression has not been given. It is sufficient to know the different styles of definition, and that aggression is a component of competition.

4.1.2 Questions about aggression

The important questions about aggression are those raised at the beginning of the chapter: Why does one animal exhibit aggressive behaviour whilst another does not? Why does one animal exhibit aggressive behaviour in some circumstances and not others? Why do some species exhibit aggressive behaviour to a much greater extent than others?

There are four types of answer to these questions, the four that were introduced in Book 1, Section 1.1.1.

1 A causal answer explains aggression in terms of the factors present (both internal and external) immediately preceding the aggressive behaviour, i.e. the factors which caused the aggression.

2 A developmental answer involves an examination of the developmental history of an animal to find reasons in its early life for the animal's aggressive behaviour as an adult.

3 A functional answer is one in which aggression is explained in terms of its consequences and how those consequences improve an animal's chances of survival and reproduction.

4 An evolutionary answer seeks to explain why a particular form of aggression has evolved in a given species, e.g. why one species engages in violent fights that can end in death, while another settles disputes by means of non-injurious displays.

All four types of answer need to be considered in response to the opening questions to give a truly comprehensive answer. There is insufficient space to do that here and, although all four types of answer are considered, this chapter focuses on causal and functional aspects. Functional and causal aspects of aggression are investigated using different experimental techniques. For example, the effect of a fight on subsequent mating success (functional approach) requires long-term observation of intact animals, whereas the effect of the hormone testosterone on fighting (causal approach) is usually assessed by removing the source of that hormone from the animal.

4.2 The causal approach to aggression

One of the central themes of this book is that behaviour depends on internal and external factors. This theme is depicted in Figure 4.1 in the form of a simple flow diagram.

Notice that the diagram includes feedback from the behaviour to both the external and the internal factors. For instance, if the behaviour involved movement then the external factors would change when the animal moved to a different location. If the behaviour involved exertion then there would be changes to the hormone and energy levels in the blood and thus the internal factors would change. Changes to the internal or external stimuli would alter the way they are integrated and so lead to further behaviour and further changes in the factors, and so on.

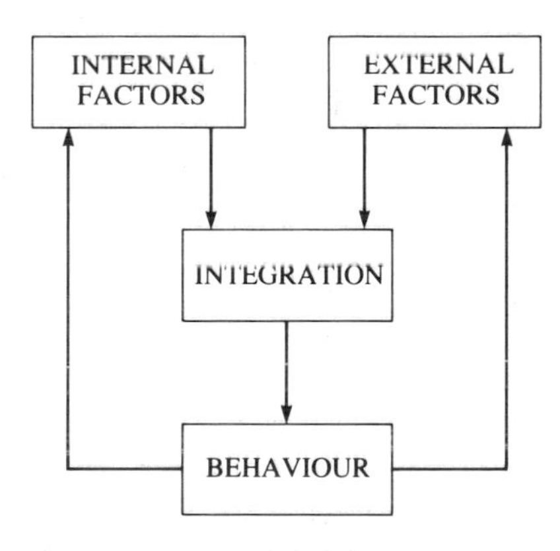

Figure 4.1 Simplified diagram showing the integration of internal and external factors to produce behaviour. In reading a flow diagram you start anywhere and move around the diagram in the direction of the arrows.

The main advantage of a flow diagram is that it allows the different components of the system to be identified, along with their modes of interaction. It is possible to use such a diagram to see how the system works. Figure 4.1 is simple and does not lead to any new insights into the topic, but it does help to conceptualize the processes underlying aggression.

4.3 Testosterone and aggression

Many hormones have been implicated in aggression, including adrenocorticotropic hormone, prolactin, oestrogen, progesterone, adrenalin and testosterone. These hormones are all found in vertebrates. Furthermore, these hormones are all found in female vertebrates, and yet most studies on hormones and aggression have been concerned with testosterone in male vertebrates. This hormone/sex combination provides the focus of this section.

There are several reasons why testosterone and aggression in the male have received more attention than other hormone/sex combinations. One reason is that in general, in a given species, males secrete far greater quantities of testosterone than females. Testosterone is also a major player in the development of sexual differences (Book 4, Chapter 4). A second reason is that males appear to engage in more frequent and conspicuous fighting than females. Thirdly, any study of behaviour in female mammals is often complicated by variations in behaviour caused by regular fluctuations in their reproductive hormones. A final reason is that in contemporary western society male aggression is seen as more of a problem than female aggression, and so the understanding and control of male aggression would be more beneficial to society.

It might appear a relatively easy matter to determine whether or not testosterone affects aggression. The first step would be to remove the principal source of the hormone, in males by removing the testes, and then examine the effect on aggression. What could be simpler?

Unfortunately it appears that the effect of this procedure (castration) and its necessary companion procedure of replacement therapy (i.e. replacing the hormone by injection) on aggression is dependent on at least four other factors. These factors are: the species, the developmental history of the animal, the experience of the animal (particularly of fighting), and the situation in which aggression is assessed. Each of these factors is now considered in turn.

4.3.1 Species differences

In many species of mammals, castration in male adults reduces the amount of fighting between them. This has been shown in field studies with voles, red deer (*Cervus elaphus*) and cats and, in the laboratory, in some strains of rats and mice. However, it is far from a universal finding. In gerbils, dogs, rhesus monkeys (*Macaca mulatta*) and people, castration does not necessarily reduce fighting.

There are wide species differences in the response to castration, depending on the normal life-style of the animal. For example, in many species of monkey, where there is a strong social order, castration has no effect on aggression: the dominance relationships that existed before castration continue after castration. For other species, castration may reduce aggression immediately or not at all. Felicity Huntingford and Angela Turner working in Glasgow, Scotland, have suggested that where males fight for direct access to females, they would only do so when ready to breed, a condition brought about by high testosterone levels. Therefore there should be a close relationship between the level of testosterone and incidence of fighting in such species. Where competition for females is indirect, by status or territory, the relationship between testosterone level and fighting frequency would be weaker. There is some support for this suggestion from seasonally breeding species such as red deer, camels and ground squirrels. In hamsters, males fight with each other to establish territories and females choose males on the basis of those territories. As the breeding season approaches the testosterone levels of the males rise and the amount of fighting they engage in decreases.

□ Explain how the hamster data also support the suggestion of Huntingford and Turner.

■ Male hamsters compete for females indirectly, through territories. Indirect competition, according to Huntingford and Turner, is compatible with the rising testosterone levels and the falling fighting frequency seen in hamsters.

4.3.2 Developmental history

In Book 4 the organizational effect of testosterone on sexually dimorphic behaviour patterns was discussed. Aggression is usually a sexually dimorphic behaviour and is often directed differently by the two sexes. Aggression between males is more frequent than that between females. For example, in many birds males attack males that seek to mate with their partners; females seem less concerned about this but do strongly defend their nests, eggs and chicks.

This sexually dimorphic aspect of aggression is organized by testosterone. Adult male mice castrated just after birth take longer to fight when injected with testosterone as adults, and require higher doses of testosterone before they do so than males castrated at later ages. On the other hand, female mice given injections of testosterone soon after birth fight more readily when treated with testosterone as adults than females which did not receive the neonatal testosterone treatment.

Chapter 4 in Book 4 described how the level of male sexual behaviour displayed by an adult is dependent on the level of testosterone that its nervous system was exposed to earlier. A similar principle applies to the level of aggressive behaviour. When adult, female mice embryos which developed between two male embryos in the uterus (womb) have a stronger testosterone-induced aggressive response to males than adult females which had developed next to just one male, or none at all. Male embryos that develop between two female embryos are relatively insensitive as adults to the aggression-promoting effects of testosterone.

In addition to the perinatal effect just described, in some species testosterone also has a pubertal effect. For instance, in mice, males that have not yet reached puberty are not attacked by adult males, nor do they attack adult males. Such pre-pubertal males may fight with each other. After puberty they respond to and promote responses in other adult males, which lead to fighting. The change caused by testosterone at puberty is sometimes referred to as an activating effect of the hormone. In such terms, at puberty, testosterone would activate those systems responsible for aggression.

4.3.3 Experience of fighting

In group-living species such as rhesus monkeys and rats, animals that are successful at fighting generally occupy positions towards the top of the dominance hierarchy and are often referred to as high ranking, while those that are unsuccessful at fighting occupy positions towards the bottom of the hierarchy and are referred to as low ranking. High ranking and low ranking animals differ in hormonal state, with high ranking animals having a higher level of testosterone than low ranking animals. Furthermore, winning a fight tends to increase the level of testosterone, whereas losing a fight tends to decrease it. A small difference in fighting ability or status may, after a series of fights, lead to a large difference in testosterone levels. Thus, differences in testosterone levels can result from, rather than cause, the differences in fighting ability. In other words, the effects of winning or losing a fight on hormone production are such as to perpetuate and exaggerate small status imbalances.

An interesting consequence of these changes in hormone production has been found in rats, mice and bonnet macaques (*Macaca radiata*). They appear to provide a method by which high ranking animals incapacitate reproductive competitors, so that high ranking animals reproduce whereas low ranking animals do not. This consequence has been called 'psychological castration', because the failure to reproduce is a product of the effect of the high ranking animal on the low ranking one.

Apart from the hormonal consequences of losing or winning fights, the animal also learns about winning and losing, an effect clearly demonstrated in rats and mice. Differences in aggressive behaviour may depend on the first few opponents an animal encounters as an adult. If its first two fights as an adult happen to be with a high-ranking conspecific, its subsequent aggressive behaviour is likely to be different from that of an animal whose first two fights were with a low-ranking conspecific. In this case one animal experiences losing badly and the other does not. In future, the loser might well react with a fear response (e.g. by freezing or escape) rather than attack. An animal that is put into a situation where it always loses the fight learns that in that situation it always loses, and so it changes its behaviour accordingly. Similarly, an animal put in a situation where it always wins, learns to win. The behaviour of these 'trained losers' and 'trained winners' differs in other situations because of their different experiences of fights.

4.3.4 Social context

The context in which animals compete affects the nature of that competition. For instance, a female (e.g. herring gull, *Larus argentatus*; pig; cat) might well remain to defend a nest containing offspring against a conspecific whereas at other times, in the absence of nest and offspring, she would flee from the same conspecific. The importance of the social context is well illustrated by the following two experiments undertaken to determine whether castration affected aggression in male Mongolian gerbils (*Meriones unguiculatus*).

The first experiment investigated the effects of olfactory bulb removal and manipulation of androgen on the marking and aggressive behaviour of male Mongolian gerbils (Lumia *et al.*,1975). Adult male gerbils were bought from a commercial supplier and kept alone at all times except during the actual tests. Each animal was tested at three different times, in the following order: (1) a pre-operation test soon after acquisition, (2) a post-operation test after either castration, or a very similar operation that did not include removal of the testes, termed a 'sham operation', and (3) after injection with either an inactive substance (oil), or testosterone. At each of the three times the animal was subjected to four tests against different intact ('normal') opponents.

The tests consisted of putting one experimental and one intact opponent together in an arena 60 × 60 cm square and recording the number of fights that occurred in 10 minutes. Each animal was only tested once on any one day.

The results for the twenty castrated animals (castrates) and eight sham treatment animals are presented in Table 4.1.

□ In this experiment, do members of the castrated group fight more or less frequently than the control group before the operation (pre-op)?

■ Although the castrate group have a slightly higher mean number of fights than the control group, the difference is not significant. The frequency of fighting in the two groups can therefore be regarded as the same.

Table 4.1 Effects of castration and testosterone injection on fighting in male gerbils.

	Mean number of fights per 10-minute trial			
	pre-op	post-op	post-injection	
			oil	testosterone
castrate group	4.1	0.8	0.3	2.7
sham group (controls)	3.4	2.9	3.0	–
significance	NS	$P < 0.01$	$P < 0.01$	

NS = no significant difference between groups
pre-op = before castration or sham treatment
post-op = after castration or sham treatment

□ Do the castrates fight more or less frequently than the sham group after the operation?

■ The castrates fight less frequently than the control group.

□ What evidence is there that the reduction in frequency of fighting following castration in this experiment was due to the loss of testosterone?

■ There are two pieces of evidence. First, removal of the principal source of testosterone, the testes, reduced the amount of fighting, whilst those operative procedures common to the sham operated group (e.g. anaesthesia, handling, etc.) did not. In other words, the sham treatment had no effect but the removal of the testes did have an effect. Second, injection of testosterone restored the level of fighting in the castrates, whereas injection of a neutral substance (oil) had no effect.

The second experiment investigated the effect of castration on social reactivity in the male Mongolian gerbil (Christensen *et al.*, 1973). Researchers bought adult male gerbils from a supplier and subsequently kept each male gerbil alone in a cage at all times, except when the actual tests took place. Eight animals were castrated and formed the castrate group, and eight were left intact to form the control group.

The tests were carried out using the apparatus shown in Figure 4.2. Animals were tested in pairs. One subject was placed in each compartment of the test cage a week before it was tested. Each subject was allowed regular access to the central area, but only ever alone. Both subjects in the apparatus were from the same group, either control or castrate. During the course of the experiment each subject was paired in the apparatus with each other subject in the same treatment group, a so-called round-robin procedure.

After the animals had lived for a week in the apparatus the actual test was carried out. The two animals were allowed access to the central area simultaneously and were observed for 10 minutes. The number of fights, and the number of fights which had to be terminated because blood was drawn were recorded. The results for each of the 28 possible pairings in each group are presented in Table 4.2 (*overleaf*).

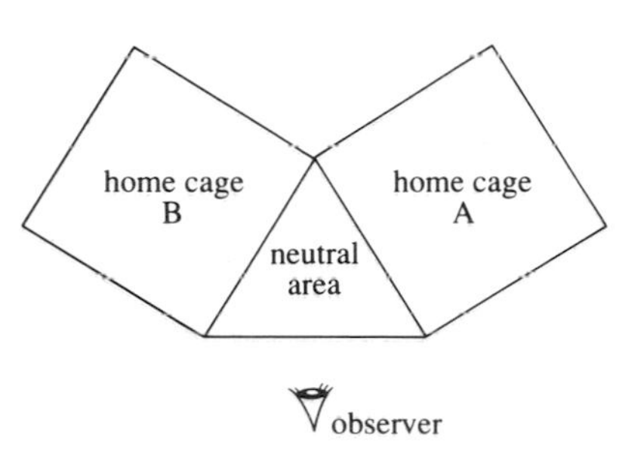

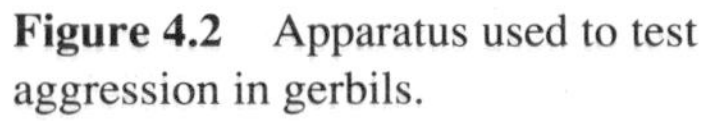

Figure 4.2 Apparatus used to test aggression in gerbils.

Table 4.2 Incidence of fighting in pairs of control and pairs of castrated male gerbils.

	controls	castrates
tests with fights	6	19
tests with no fights	22	9
tests terminated	3	17
tests not terminated	25	11

□ From the table deduce whether the castrates or the controls were: (1) more likely to fight, and (2) more likely to draw blood.

■ Pairs of castrates were more likely to fight (19 out of 28 compared with 6 out of 28) and more likely to draw blood (17 out of 28 compared with 3 out of 28) than pairs of intact controls.

□ Consider the results of the two experiments just described. Does testosterone affect fighting in gerbils?

■ Yes it does. In both experiments 1 and 2 castration altered the frequency of fighting of the gerbils, but not in any predictable direction.

Castration, then, had diametrically opposite effects in the two experiments. In the first experiment castration decreased fighting frequency and in the second experiment castration increased fighting frequency. The main difference between these two studies is the way in which fighting was elicited. The important point to emerge from considering these two experiments is that the effect of testosterone on aggressive behaviour is dependent on the context in which aggressive behaviour is performed.

It is clear that integration involves rather more than one internal factor, testosterone, and one external factor, another adult male. The relationship between testosterone and aggression in the adult male can perhaps be best described as variable, with the variation arising from the interplay of the four factors considered above.

There is one final aspect of these studies which cannot be ignored, and that is what behaviour patterns should be selected as being a measure of aggression? This question brings the chapter right back to the beginning where definitions were considered. Most laboratory studies of aggression have defined aggression by form, usually using attack or fighting as their measure of aggression. It is arguable that this is too narrow a view of aggression. It certainly excludes a major facet of aggression, namely the acquisition of a resource. For instance, in the second experiment discussed above, if, instead of fighting, possession of the small neutral area had been used as the measure of aggression the results might have been different. Indeed, fighting frequency can sometimes be misleading. In well-established troops of rhesus monkeys, for instance, the highest ranking male is rarely involved in fights because he is simply avoided by other males. Lower ranking males frequently fight amongst themselves. For purposes of the present discussion the issue is whether the highest ranking male can be considered the most aggressive male despite fighting only infrequently because he has first access to all resources? There is no simple answer to this question, but the question itself highlights the importance of considering the social context when measuring aggression.

4.3.5 How does testosterone exert its effects?

In many species testosterone does affect aggression though, as was explained in the previous section, the exact relationship between testosterone and aggression in the adult male is variable, and not subject to generalization. By what mechanism then does testosterone affect aggression? This question is a specific formulation of the more general question: How can a hormone affect behaviour?

Firstly, there are the organizational effects of testosterone. Organizational effects were described in Book 4, Chapter 4, and consist of effects upon the structures that will later play a role in behaviour. Cells (particularly neurons in the CNS) might, or might not, be responsive to testosterone or its metabolites, oestrogen and dihydrotestosterone. Those cells that are responsive might survive rather than die, or might start to divide and differentiate. Such testosterone-sensitive cells form part of the future neural basis of aggressive and sexual behaviour. Secondly, there are the later effects of testosterone upon the animal. Structures already organized can be sensitized by testosterone later on, making aggression more likely in a given situation. Thus the sensitization alters the motivation. In some species, this motivational effect will be revealed in a variation in aggressive tendency as a function of testosterone level.

Testosterone can also affect aggression in an indirect way. For instance, it can affect body size by altering growth rates, body shape through the growth of new structures, sensitivity of sense organs, coloration by the production of new pigment, or smell by the production of new odours. All of these changes could influence the way an animal responds to other animals, and the way other animals respond to it.

An example of these indirect effects of testosterone on aggression is seen in the stickleback. Absence of testosterone in the adult male stickleback prevents it from synthesizing the glue necessary to build its nest. Absence of the nest means there is nothing for the fish to defend, and so it is not aggressive towards male intruders. Another example of an indirect effect is the production and detection of odours in the urine of mice. When two strange, intact, male mice meet, a fight normally ensues. Castrated male mice produce urine which lacks a particular odour normally associated with adult male mice and neither attack, nor are attacked by, intact males. Castrated male mice painted with the urine of intact males are attacked by intact males, showing that the urine of intact males contains an odour which influences aggression. Furthermore, castrates fail to distinguish the urine of intact males from that of castrates, a task easily accomplished by intact males. Thus, testosterone is required not only to produce a particular odour, but also to detect it, and the odour affects aggression.

4.3.6 Variations in the level of testosterone

As with many other hormones in an adult animal, the amount of testosterone in the blood, the circulating level, varies from time to time. The variations occur on a number of different timescales, most of which you have already met.

Seasonal variations ensure that animals return to breeding condition at the appropriate time of year. Day length or temperature may be the external factors which trigger a seasonal rise in testosterone level (Section 3.9.1). There is evidence that the presence of a receptive female can lead to a rise in testosterone levels in adult male rats and mice. This particular change in testosterone level is very quick and quite unlike the longer term reduction in testosterone level seen between a subordinate male and a dominant male. A third variation in testosterone level arises because testosterone is secreted in short bursts (as with some other hormones; see also Chapter 3) rather than continuously.

All of these variations are superimposed on one another and a direct correlation between testosterone level and behaviour is therefore difficult to make.

4.3.7 Men, testosterone and aggression

There are some similarities between the effects of testosterone on people and those described above. For instance, developmental effects of testosterone (i.e. organizing and activating effects) have been demonstrated in people (see Chapter 4 in Book 4), and there are also short-term changes in testosterone level in men dependant upon recent competitive experiences (e.g. playing tennis or wrestling). There are also some tantalizing suggestions of a relationship between testosterone and aggression (here defined as criminal conviction) in adult men. (Behaviour patterns as diverse as verbal abuse, and conviction for violent crime and rape, are frequently used as examples of aggressive behaviour in discussions of human aggression. Hand to hand fighting would best accord with the definition of aggression by form used in this chapter, though it is rarely assessed.) Men convicted of crimes have higher levels of testosterone on average than men with no criminal convictions (Figure 4.3a). Men convicted of violent crime have higher levels of testosterone than men convicted of other crimes (Figure 4.3b) and this difference probably accounts for the difference between criminals and non-criminals, since there is no difference between men with no criminal convictions and men convicted of non-violent crimes.

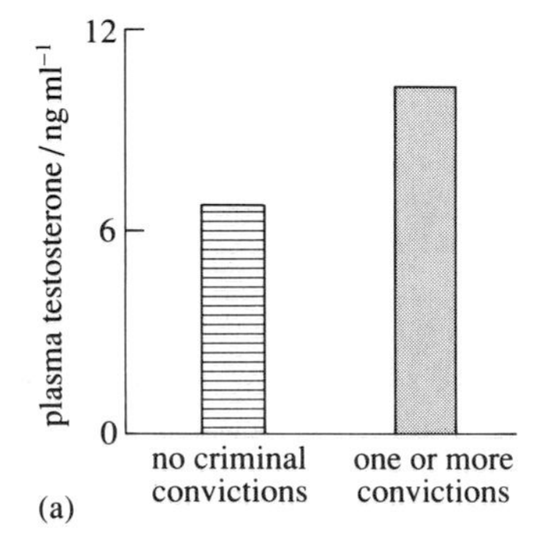

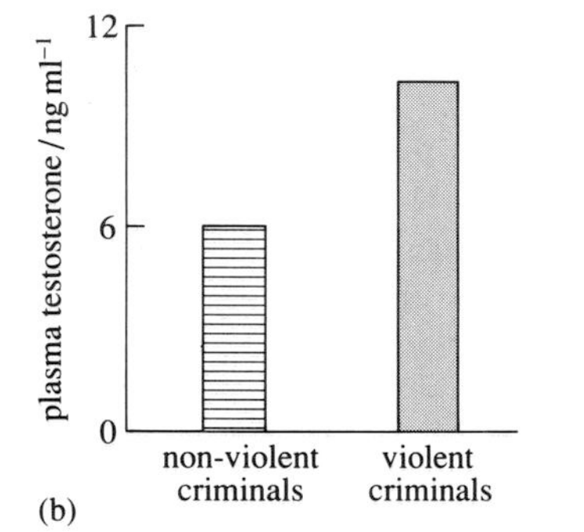

Figure 4.3 Level of testosterone circulating in the blood (ng ml^{-1}): (a) in men with and without criminal convictions, and (b) in non-violent and violent criminals.

Such clear data are the exception, though even here other results throw doubt on what the data mean.

□ Does a positive correlation between aggression and testosterone prove that testosterone causes aggression?

■ No. It is possible that it is a violent life-style that leads to a rise in testosterone level.

If testosterone is having some effect on criminality, as is suggested by the data, then reducing the level of testosterone ought to reduce the rate of crime. However, when drugs are used to reduce the level of testosterone in criminals and sex offenders, there is little effect on their rate of re-offending.

In general, the studies on testosterone and human aggression reach the conclusion that testosterone is involved, but it is not of over-riding importance.

Summary of Section 4.3

The main message of this section is that the relationship between testosterone and aggression is not a simple one that can be encapsulated within a generally applicable statement. First, there are species differences. Second, both the internal factors and the external stimuli to which an individual has been exposed alter the relationship between testosterone and aggression for that individual. Thirdly, the situation in which aggression is assessed has a profound influence on the aggressive behaviour of an individual. A final point is that, where a relationship can be demonstrated between testosterone and aggression, testosterone may act on some peripheral structure to alter scent production or shape, say, rather than on the CNS.

4.4 The brain and aggression

During an encounter, an animal perceives a potential opponent in terms of its species, gender, size, previous encounters, and reproductive state. On the basis of this perception, it decides whether and how to fight. It puts these decisions into effect by orchestrating the movements that constitute aggressive behaviour, with these comparisons and decision processes occurring in the brain. This section addresses the question of whether there are particular parts of the brain with particular responsibility for aggressive behaviour.

4.4.1 Output and input

Output

By and large, specific behaviour patterns performed by an animal do not require specific sets of muscles dedicated to that particular behaviour. Fighting in anemones (Book 2, Section 7.2.4), for instance, was the result of the particular *sequence* of contraction of the same muscles used in protective contraction. The behaviour is determined by the sequence of muscles contracted, not by special muscles. Biting during fighting in any species uses the same muscles as biting during feeding. Head butting in rams and the use of antlers by stags require the same leg and neck muscles used for running and grazing, respectively. The few behaviour patterns which do have dedicated muscles associated with them also have specific motor nuclei in the brain.

You may recall from Book 4 that penile erection in the rat is dependant on dedicated muscles that are innervated by axons originating in a specific motor nucleus.

□ Given that aggressive behaviour does not rely on dedicated sets of muscles, what might you deduce about brain structure and the control of aggressive behaviour?

■ A reasonable deduction might be that there is no brain structure dedicated to aggressive motor output.

Put another way, brain structures involved in aggression influence: (1) the motor cortex, which in turn influences the patterns of muscle movements, (2) the necessary physiological changes which accompany aggression, and (3) which external stimuli are selected as targets. However, the structures have no direct motor output themselves. Thus the motor output leading to aggressive behaviour is via pathways and systems used in many other behaviour patterns.

Input

Whether an animal is aggressive or not depends in part on a whole host of factors, some of which have been considered already in this chapter. The numerous internal and external factors are integrated in some way and, if the balance is appropriate, aggression results.

It was noted in the previous section that many rodents (rats, gerbils, etc.) are highly responsive to smell. If an adult male mouse is unable to smell (i.e. it is anosmic), it does not attack other adult male mice, although it is frequently attacked by them. A mouse can be made anosmic by removal of the olfactory bulbs. The olfactory bulbs do not control aggression, and yet their removal alters an animal's propensity to fight by altering its sensitivity to particular stimuli. Similarly, there is a pathway from the olfactory lobes to the amygdala, so that removal of the amygdala often, though not always, results in a docile, non-aggressive animal. One of many possible explanations for the results of amygdalectomy (removal of the amygdala) is that olfactory stimulation is no longer interpreted in the same way and the animal is rendered docile. Thus, amygdalectomy may alter the propensity to fight by changing the way in which stimuli are integrated.

These are two principal ways in which stimulating or lesioning particular brain regions can change the balance between the various internal and external factors. One way is if the lesion, for example, simulates a change in the intensity of a sensory modality, like the anosmic rat previously mentioned. The second way is by stimulating or lesioning a nucleus, or part of it, that is involved in integration, as with amygdalectomy.

Thus Figure 4.1 can be further modified and is shown for illustrative purposes only as Figure 4.4 (do not make an effort to memorize the details of this figure). Both the limbic system, of which the amygdala is a part, and the hypothalamus are shown to receive internal and external stimuli, and they both have an integrative function. Output is via the motor system, the ANS and the endocrine system (Chapter 5). The figure is beginning to get unwieldy and will not be complicated further, although the fact that 'limbic integrations' is merely a box which reveals nothing of the intricate connections between the various structures of the limbic system attests to the actual simplicity of the figure compared to the real brain.

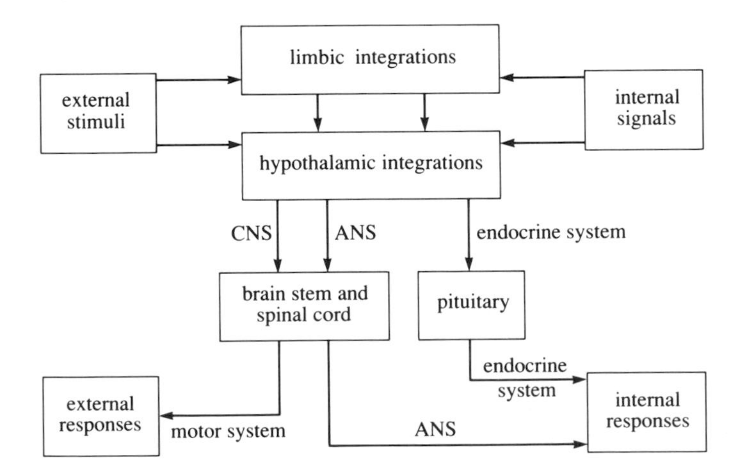

Figure 4.4 A further modification of the flow diagram presented in Figure 4.1.

The next section examines the involvement of one particular area of the brain, the hypothalamus, in aggressive behaviour. Section 4.4.3 considers other brain areas implicated in aggressive behaviour.

4.4.2 The hypothalamus

Numerous studies of species as diverse as sunfish, lizards, ducks, mice and monkeys have suggested that the hypothalamus is closely involved in the control of aggression. The study reported here, by David Adams (1971) in the United States, is fairly typical, both in terms of procedure and results. It involved lesions to the hypothalamus of rats.

Adams used two different lesions: (1) a lesion to the medial hypothalamus, and (2) lesions to the lateral hypothalami (there is a left and a right lateral hypothalamus, see Figure 4.5). Six rats were subjected to each type of lesion and each of the 12 rats was tested for aggression in two situations. The first aggression test was an electric foot-shock test. (This obnoxious, and now rarely used test involves pairing two animals in a small arena with a metal grid floor. Pairs of control rats attack each other when an electric current is passed through the floor.) The measure of behaviour used, called defensive behaviour by Adams, was defined as both animals on their hind legs, forepaw to forepaw, as if boxing. The second test involved placing an intact rat in the home cage of the lesioned animal. The measure used was the number of minutes out of ten in which fighting occurred. This behaviour was referred to by Adams as territorial behaviour. Adams' results are presented in Figure 4.5.

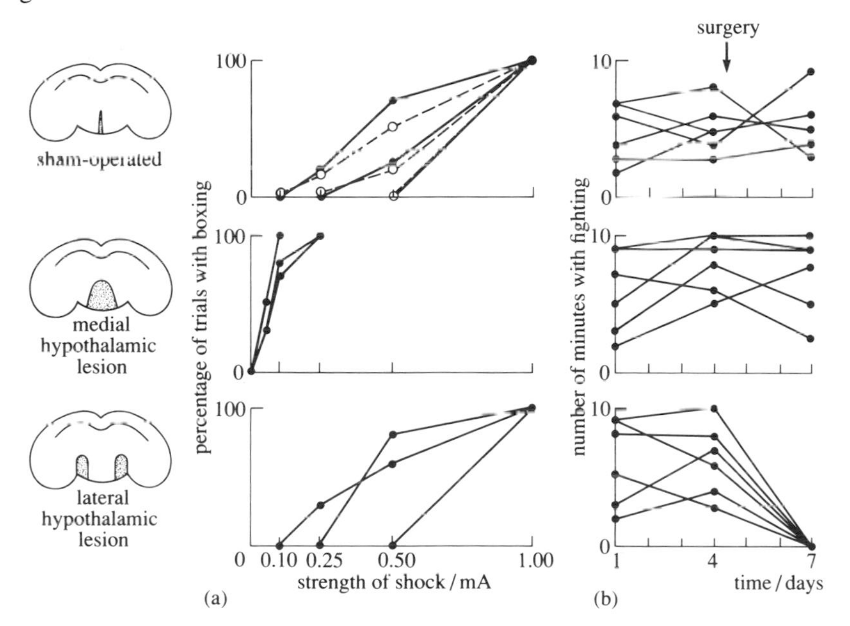

Figure 4.5 Effects of hypothalamic lesions in two different places on two behaviour patterns in rats. The position of the lesion is shown on the left; sham-operated control rats are shown at the top. (a) plots the amount of electricity delivered to the feet of a pair of rats (measured in milliamps, mA) against the percentage of trials in which the behaviour defined as defensive behaviour by Adams ('boxing') occurred. The dotted lines indicate results prior to the sham-operated test. (b) plots time (in days) against the number of minutes in which attack occurred in a 10-minute test. Each separate line on a graph represents a different animal.

□ Look at Figure 4.5a. Which lesion (medial or lateral) had little or no effect on defensive behaviour in response to electric foot-shock?

■ The bottom graph is similar to that for the controls (the top graph). Thus lesions to the lateral hypothalamus have little or no effect on defensive behaviour.

□ What effect do lesions to the medial hypothalamus have on responsiveness to electric foot-shock?

■ This lesion makes the animals responsive to lower levels of shock than control animals. Therefore these lesions make the animals more responsive.

Now look at Figure 4.5b. (Note that two trials were given before the surgery was performed, one on each of days 1 and 4, and one trial was given after surgery, on day 7.)

□ Which lesion had little or no effect on attack behaviour in response to an intruder?

■ The number of minutes in which attack occurred was very similar before and after surgery for both the controls and for animals with lesions to the medial hypothalamus.

□ What effect did the other lesion have on attack behaviour?

■ The number of minutes in which attack was recorded was much reduced in animals with lesions to the lateral hypothalami. Thus lesions to this nucleus reduce attack behaviour.

□ On the basis of these data would you say that the hypothalamus is, or is not, involved in aggressive behaviour?

■ The hypothalamus is involved in aggressive behaviour.

This study demonstrates two important points. Firstly, a particular lesion can affect two different measures of behaviour, both of which have been regarded as measures of aggressive behaviour, in different ways.

□ What evidence is there for this statement?

■ Lesions to the lateral hypothalami have no effect on defensive behaviour but reduce territorial fighting. Lesions to the medial hypothalamus heighten defensive behaviour but have no effect on territorial behaviour.

At issue here is the necessity of knowing the actual behaviour patterns observed rather than relying on a convenient, but misleading, label like aggression. If an experiment purports to show an increase in aggressive behaviour, it is crucial to know what measure of aggressive behaviour was used.

The second point is that different lesions within the same structure (the hypothalamus) can have different effects, even if the same measure of behaviour is used. The effect on territorial behaviour of lesioning the medial hypothalamus is different from that of lesioning the lateral hypothalami.

Another type of study, using the complementary technique of electrical stimulation of the brain (Book 2, Chapter 10), makes this point even more clearly. An example of this type of study was carried out in Holland (Kruk *et al.*, 1983). The researchers placed electrodes in the hypothalami of 270 rats. The aggressive behaviour of these rats in response to electrical stimulation of the brain was then assessed by placing them individually in a box in the presence of an unfamiliar rat which had no electrodes in its brain (a control rat). When the rat was artificially stimulated by turning on the electric current to its electrodes, one of three behaviour patterns was recorded: (1) no response, (2) biting, where the stimulated rat bit the control rat without preliminary fighting (sometimes referred to as a quiet biting attack), and (3) attack, where the stimulated rat attacked and fought the other rat. The amount of electricity necessary to elicit any behaviour (i.e. a threshold) was established by using *ascending* and *descending* trials. First, the amount of electricity (current) was set at a low level and was increased in stages until a behaviour pattern was reliably performed when the current was switched on. (Note: the current was not altered whilst the electrodes were switched on.) Then the current was set at a level higher than the threshold revealed by the ascending trial and then decreased until the behaviour was no longer performed when the current was switched on. This procedure of increasing and decreasing the current was repeated several times and a threshold current (i.e. the minimum current required to produce a behaviour pattern) was determined.

In 150 of 411 electrode placements no response was recorded. A further 67 electrodes had a weak or erratic behavioural effect, leaving 194 electrodes with clear positive effects.

The behavioural response elicited by the lowest current was biting in 73 placements and attack in 121 placements. Higher currents delivered to the 73 placements elicited attack.

Menno Kruk and his colleagues then took sections of the rats' brains to identify the precise location of the electrode tip and hence the point of stimulation. Stimulation of parts of the lateral hypothalamus, parts of the medial hypothalamus, and parts of the anterior hypothalamus all resulted in attack. However, some electrodes in each of these anatomically defined areas (distinguished by cell density and cell shape) did not elicit attack when stimulated.

The principal conclusion from this study was that the area which, when stimulated, resulted in aggressive behaviour does not coincide with classical anatomical distinctions such as medial hypothalamus or lateral hypothalamus. Thus different anatomical structures within the hypothalamus are functionally sub-divided. Small discrete areas of one structure within the hypothalamus may be involved in one neural system, while adjacent areas are involved in other systems. Changing the current delivered to an electrode can alter the behavioural effects of that electrode by increasing the stimulation to the same neurons, and by stimulating more distant neurons, which may lie in any direction from the electrode tip, and may be part of another neural system.

The hypothalamus is a small structure, roughly 3 × 3 × 3mm in the rat. Some parts of it are clearly involved in aggression. Yet exactly what its role in aggression is and which neural elements are involved must await even more detailed study.

4.4.3 The amygdala and septum

Two other areas of the brain, the amygdala and the septum, have also been implicated in aggressive behaviour. Generally, removal of the amygdala (in rats, cats and monkeys) renders the animal tame, docile and difficult to provoke. Stimulation of the amygdala has been reported to result in quiet biting attack, and in 'affective defense' in rats and cats. 'Affective defense' is behaviour typically seen when a cat is cornered (ears flattened onto its head, teeth bared, eyes part closed), though the behaviour occurs when the amygdala is stimulated and no other animal is present. Stimulation of certain locations within the amygdala elicits one of these behaviour patterns, whilst stimulation of other locations within the amygdala elicits the other behaviour pattern. The amygdala then, like the hypothalamus, has discrete regions within it which do different things.

Removal of the amygdala was used in the 1950s and 60s to control violent men. The results were far from conclusive, with some operations being successful but many being unsuccessful. The removal had little or no effect on violence in the majority of the men operated on. Where the removal did reduce violence, such reduction appeared to be short-lived, with the men returning to their pre-operative behaviour patterns in 6–12 months.

The septum, and its involvement with aggression, has been less well studied than that of either the hypothalamus or amygdala. Removal of the septum makes rats easily provoked, difficult to handle, and highly responsive to the electric foot-shock test. Stimulation of the septum completely suppresses fighting in hamsters. The septum, like the hypothalamus and the amygdala, is a complex structure, and its complete removal in rats and guinea-pigs is often accompanied by increased general activity and gregariousness. These general changes make the interpretation of experiments in terms of aggression that much more complicated.

The studies described above and in the previous section suggest that the hypothalamus, amygdala and septum have some involvement in aggressive behaviour. Exactly what that involvement is and what part these regions of the brain play in the decision processes of the animal are questions that await further study.

4.4.4 The theoretical framework

Many of the advances in the study of neural structures underlying behaviour have resulted from improved technology, e.g. better electrodes, better tracing techniques, improvements that no doubt will continue. The theoretical framework of these studies has also changed.

Early studies were based on a theoretical premise that particular brain nuclei control particular behaviours. By removing a given nucleus, so a specific behaviour could be prevented or enhanced. The picture of the brain that now emerges is of an extremely intricate structure with different regions of the same anatomically discrete area doing different things. The subtlety of this picture is quite unlike that envisaged in previous decades, where it was the anatomically distinct areas that were assumed to be doing different things. Furthermore, the brain, or rather the activity of the neurons of which it is composed, is dynamic; the activity of the neurons is incessant. It is not possible to remove one small part of the brain and observe how that small part functions in isolation because each small part is utterly dependant on other small parts. It is the interaction between those small parts that is the very essence of how the brain functions. When a small part of the brain is removed, e.g. by lesioning, observing how the animal then behaves reveals how the rest of the brain functions in the absence of that small part. Similarly, observing how an animal behaves when a part of its brain is stimulated reveals how the brain functions when that stimulation is added to its normal activity. It is rather too easy to be deceived by focusing simply upon the most obvious effects of a lesion or stimulation of a circumscribed brain region. The more subtle issue is how the rest of the brain copes with the alteration in normal function of the circumscribed area.

A related point is that producing an effect on behaviour by lesioning or stimulating one area does not mean that other areas are not involved in that effect. It is clear that the hypothalamus, amygdala and the septum are involved in aggression, but there may well be other areas that also contribute to this complex behaviour.

Finally, it should not be assumed that an area is solely concerned with a particular behaviour. Although this appears to be true in some cases (e.g. SNB and penile erection) it needs to be established rather than assumed. For instance, quite apart from an association with aggression, a lot of attention is now focused on the role of the amygdala in memory, and the hypothalamus is involved in sexual behaviour and homeostasis. Here again, as with all experimental studies of behaviour, the external stimuli made available to an animal restrict the kinds of behaviour that the animal is likely to perform. If the study is focusing on aggression then the experimenter will provide an experimental set-up conducive to aggressive behaviour, and the effects of a particular manipulation on sexual behaviour or nest building or whatever will be missed. This simplified approach is necessary and correct to gain initial insight, but its limitations should be recognized.

Having now considered some causal aspects of aggression, the remaining sections deal with functional and evolutionary aspects of aggression.

4.5 The function of aggression

For the rest of this chapter, aggression is defined by consequence, and is that behaviour which occurs when two animals compete for an identifiable resource. We begin with a review of the concept of fitness that you met in Book 1, Chapter 4, and we then look at the behaviour patterns used when pairs of wolves compete for a resource. This examination of behaviour patterns raises questions about aggression which subsequent sections of the chapter address.

4.5.1 A review of fitness

Fitness is a measure of the ability of an animal to produce viable offspring which themselves reproduce. It is a measure of life-time reproductive success relative to that of other members of the population (and as such is often called relative fitness, Book 1, Section 4.3.4). It is rarely possible to study life-time reproductive success and, thus, components of fitness (e.g. number of eggs laid, number of matings, etc.) are usually studied to establish their contribution to fitness. Each component can be assessed in terms of whether it appears to increase or decrease fitness, but it is the total effect of all these components that actually matters.

The function of aggression that will be presented here is based on the following line of argument:

1. The relative fitness of an individual depends on its ability to make both somatic and reproductive effort.
2. To sustain somatic and reproductive effort, animals require resources (e.g. food, shelter, mates, territory, etc.)
3. Resources are often in short supply and are, therefore, the subject of competition between individuals.
4. Aggressive behaviour, if successful, enables individuals to acquire resources at the expense of others, and tends to enhance a component of fitness and, it is assumed, relative fitness.

You are now in a position to consider aggression from a functional perspective. Aggression is behaviour that occurs when two animals compete for an identifiable resource. The importance of the resource is that it enhances a component of fitness. This concept is easiest to understand in the context of two males competing for a female, which is often the example studied. The male that competes successfully has the opportunity (as long as the female allows) to sire offspring and hence increase his fitness. The unsuccessful male is less fit than the successful male. As you will see, the measurement of components of fitness is a major aspect of modern ethological analyses of aggression. To begin with though, there are some important features of aggressive behaviour that require explanation.

4.5.2 Behaviour during direct competition

Aggressive behaviour is an enormously heterogeneous category of behaviour patterns, and it can be highly misleading to make generalizations about what animals do when competing. No species can be said to show aggressive behaviour that is typical or representative of a majority of species, because species differ in body shape and weapons. In order to make some specific points though, the aggressive behaviour of the wolf (*Canis lupus*) is described.

Aggression in wolves

Aggressive behaviour between wolves can be divided into three stages:

Stage 1—Threat Once two wolves have identified each other as potential rivals, they show very characteristic stereotyped behaviour, in which they erect the fur on the nape of the neck, flatten their ears, partially close their eyes, retract the upper lip, and growl (Figure 4.6).

Figure 4.6 A wolf in threat posture.

Many species use distinctive behaviour patterns such as these immediately prior to fighting. Fighting does not always ensue, but it does not normally occur without these preliminary activities. The form of these behaviour patterns varies enormously from species to species. Animals might produce a noise or a pressure wave (in water), secrete a pheromone or use particular postures. Because of their close association with fighting, these patterns of behaviour are referred to as **threat** or **threat displays**. A common feature of threat displays is that they involve an apparent increase in body size; in wolves this is achieved by the raising of the fur. Another common feature is the exposure of potentially dangerous weapons; in wolves, threat involves baring of the teeth.

The competition may cease at this point with one wolf backing off or fleeing. If not, stage 2 begins.

Stage 2—Fighting If both wolves continue with their threat displays and approach each other, fighting ensues. The animals attempt to bite each other and may succeed in doing so.

When watching animals fight, it is difficult to escape the conclusion that they are trying to injure one another. Indeed there is a considerable catalogue of injury, severe injury and death resulting from fights in such species as slugs, ants, spiders, frogs, rats, the musk ox (*Ovibos maschatus*), the cheetah (*Acinonyx jubatus*) and chimpanzees.

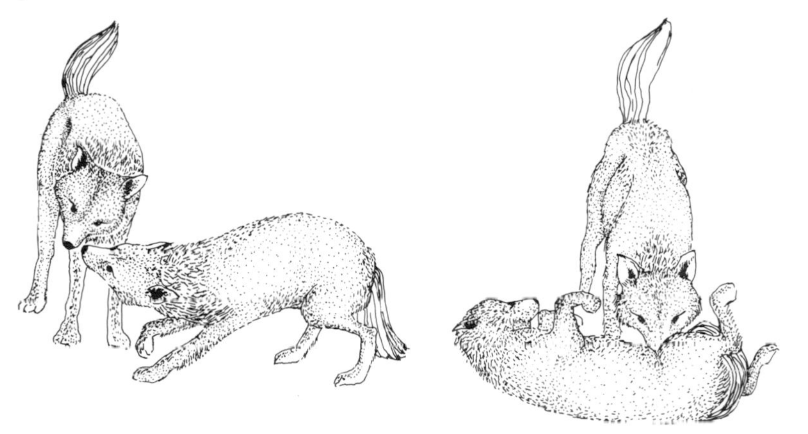

Figure 4.7 Two kinds of submissive posture shown by wolves. *Left:* one wolf mimics juvenile food-begging behaviour. *Right:* passive submission by rolling on the back.

Stage 3—Submission Fights end by one wolf running away or, more frequently, by one wolf adopting a characteristic **submissive posture**, which is opposite in form to that of the threat display. Thus, the fur is flattened, the head and tail are lowered, the legs bend, the lips are closed and growling ceases (Figure 4.7). Alternatively, it may roll over on to its back. Either of these postures stops the fight. The other wolf ceases fighting and usually starts to investigate the submissive wolf by sniffing it.

Many species adopt characteristic postures or stances which apparently stop a fight or inhibit the rival. Such submissive postures are also referred to as **appeasement displays**. Charles Darwin drew attention to the fact that submissive displays are often diametrically opposite in form to threat displays within a species (Figure 4.8), and referred to this as the principle of antithesis.

A dispute between two individuals over some contested resource typically consists of three phases: threat, fighting, and submission by the loser. However, there is much variation within this general pattern, for example, many disputes do not go further than threat, and others may end without any obvious winner or loser emerging. Animals have been described as behaving as if they were following a set

Figure 4.8 Darwin's principle of antithesis illustrated by a dog in submissive (*top*) and threat (*bottom*) postures.

of rules or conventions. The general convention used by many animals, such as wolves, seems to be: settle disputes by means of displays rather than by engaging in damaging fights. Many animals appear to behave in this conventional way, although there are some extremely violent species which do not employ threat or appeasement displays; however, these are comparatively rare. Although the catch phrase 'the survival of the fittest' conjures up images of animals perpetually fighting tooth and claw, this image is far from the truth. For ethologists, the task is to explain why animals use conventional forms of aggression such as threat and submission? How did such behaviour evolve, and what is its function?

4.5.3 The evolution of aggression

One of the earliest attempts to explain the evolution of aggressive behaviour was made by one of the founders of ethology, Konrad Lorenz, working pre- and post-World War II, in Seewiesen in Austria. According to Lorenz, the significance of the use of conventions (conventional behaviour) during disputes is that it ensures that stronger animals do not injure or kill their weaker, and often younger, rivals. Consequently, the species does not lose its reserve of growing animals. In other words, the behaviour is explained in terms of its adaptive value to the species.

□ Can you think of any reason why this sort of explanation might be incompatible with the theory of natural selection?

■ It explains the evolution of conventional aggression in terms of its adaptive value to the *species*, whereas natural selection theory is expressed in terms of adaptive value to the *individual*.

The kind of evolutionary argument in which the species, rather than the individual, is seen as the unit that benefits from the possession of a characteristic is called group selection. It is very different from evolutionary explanations that are expressed in terms of benefits accruing to individuals, which are in terms of individual selection. Group selection only works in extremely limited circumstances (see Book 1, Section 4.3.7). Couched in terms of individual selection, the argument is that conventional behaviour will evolve if it is more advantageous for an individual to behave conventionally than to behave unconventionally. The question that must be answered is, why might this be so?

Individuals tend to participate in many disputes over resources during their lives. In any dispute there will usually be a loser and a winner, and any individual is likely to lose some contests and win others. If it can emerge from those contests that it loses undamaged, it is more likely to win some of its subsequent disputes than it would if it gets wounded or exhausted when it loses. To put it another way, it is to the benefit of any individual to minimize the costs of any one fight because it will thereby increase its potential benefits from subsequent fights. This, in its normal form, is the argument for the evolution of conventional forms of aggression, expressed in terms of individual selection.

This line of argument may satisfactorily explain why it is in the best interests of an animal that is likely to lose a fight to bring it to an end quickly by performing a submissive action. How, though, do we explain the behaviour of the winner in such a dispute? Why should the winner respond to its opponent's submissive behaviour by stopping the fight? Why does it not try to wound or kill its opponent?

Again, the argument is that it is not in the best interests of the winner to prolong a fight that it has already effectively won. Not only would a prolonged fight waste time and energy, but it would also involve further risk of injury. The opponent, though defeated, may still use its weapons.

Extending this line of argument, conventional aggression could serve, not only to shorten fights, but also to fulfil another important function. Conventional aggression could enable the participants to *assess* each other before engaging in a fight. The threat display could be a means by which animals judge the size, strength and condition of a rival.

In the following sections some experimental meat is put on these theoretical bones.

Summary of Section 4.5

The great majority of animals show forms of aggressive behaviour in which little or no harm is inflicted on the participants. Instead, disputes are settled by conventional forms of behaviour involving threat displays or submissive displays. Conventional aggression can be explained in terms of individual selection, and provides a mechanism by which contests can be settled quickly and risky fights avoided.

4.6 The costs and benefits of aggression

The explanation for the evolution of conventional aggression in terms of individual selection is based on the argument that it is not to the advantage of an individual to continue a contest beyond a certain stage. This stage is that at which it is apparent to both participants which of them is the stronger. Beyond this point, both animals may be at a disadvantage in terms of wasted time, energy expended and possible injury, but the winner cannot increase the advantage it has already gained. Another way to couch this kind of argument is to express disadvantages and advantages in terms of costs and benefits. Ideally, costs and benefits would be measured in terms of reductions and increases in an animal's relative fitness but, as was pointed out in Section 4.5.1, in practice ethologists have to content themselves with measuring variations in components of fitness. These could be measures such as quantity of food gained, number of matings achieved, or size of territory defended. The following section considers some studies in which attempts have been made to measure the costs and benefits of aggression.

4.6.1 The costs of aggression

The greatest cost an animal can incur through aggression is death, because after death there is no possibility of reproducing! Death does not necessarily occur during the aggressive interaction itself. An animal that is severely wounded may bleed to death later or may become infected. A more likely occurrence is that, as a direct result of wounds inflicted during a fight, a wounded animal is taken by a predator which finds it easier prey than healthy animals. Wounded animals may also become weaker as a result of a reduced ability to find or catch their food. Even if an animal is not so handicapped that it cannot meet the normal rigours of everyday life, it is likely that its capacity to engage in future aggressive interactions will be reduced by any wound.

Another cost of fighting is loss of time. Time spent on aggressive behaviour is time which cannot be devoted to other activities, such as feeding. An animal such as a great tit has, on average, to achieve an input of one insect about every half minute of the day to maintain itself in good condition. Clearly, in such circumstances, prolonged aggression could be very costly. Aggression, and in particular, fighting, also use up a lot of energy, and a loss of competitive ability through exhaustion, however brief, may represent a substantial cost. Finally, when two animals become very preoccupied in a dispute over a resource, another animal may take the resource while it is unattended. This sometimes occurs when red deer stags are fighting for females (hinds). A stag that has already herded together a group of hinds may lose them to a third stag when he gets involved in a prolonged contest with a rival stag.

Some animals sustain considerable costs as a consequence of their aggressive behaviour. In both red deer and elephant seals (*Mirounga angustirostris*), males engage in prolonged and often damaging fights to establish a group of several females. By the end of a breeding season, the successful males are so weakened by their efforts that they suffer a significantly increased probability of dying before the next breeding season. In both these species, the life expectancy of males is considerably less than that of females.

In principle, it should be possible to measure the costs of fighting.

□ How might the costs of fighting be measured?

■ By comparing some measure of fitness in those animals that fight a lot and those that fight only a little. Ideally, the measure would be in terms of life-time reproductive success.

In practice such measurement is usually very difficult because (a) it is often not possible to identify which individuals are the parents of which offspring, and (b) it would need to be continued for the life-time of many individuals.

One study in which this kind of evidence has been obtained is that by Tim Clutton-Brock and his co-workers, who are studying red deer on the island of Rhum off the west coast of Scotland. In this study, red deer have been observed in meticulous detail throughout many years, and particularly during the autumnal rut (mating season), and it has been possible to record which animals have been the parents of most of the calves.

This study has revealed (Clutton-Brock *et al.*, 1979) that all stags sustain some sort of injury during their lifetime as a result of fighting during the rut. Of these injuries, about 20–30% are permanent, involving such wounds as the loss of an eye, or leg damage leading to permanent lameness. This figure is probably an underestimate because it does not include invisible internal injuries. Nor, probably, does it represent the real cost of fighting that would be incurred under natural conditions. Although Rhum is a nature reserve, it contains none of the predators, such as wolves, which would naturally eliminate wounded deer.

A good example of the cost incurred by an individual stag as a result of fighting is shown in Figure 4.9. This shows the number of hinds grouped together by a 7-year-old stag called Congal during the 1974 rut. In the middle of this rut he sustained a severe leg injury, possibly a torn ligament, during a fight.

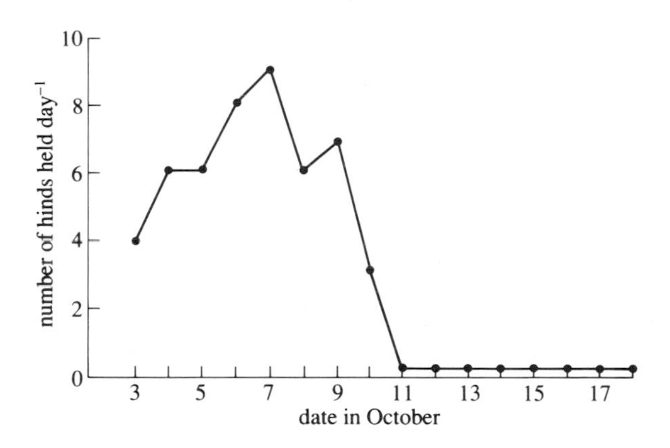

Figure 4.9 The number of hinds grouped together by a single stag called Congal, during the 1974 rut. He received a severe leg wound on 10 October.

□ What was the apparent effect of Congal's injury on his ability to group hinds together?

■ It was totally lost. Before his injury Congal was able to keep an average of seven hinds grouped together each day. After the injury he could group together none.

□ What assumption must be made to infer that Congal's injury altered his relative fitness?

■ That he would have sired offspring with the hinds that he was keeping in a group, had he not been injured.

Hinds become sexually receptive in the latter half of October and each hind produces one calf. Thus, on the basis of his performance prior to his injury, Congal could have been expected to sire about seven calves in 1974; after his injury he could not have sired any.

This does not mean that Congal's overall lifetime fitness was zero, because red deer stags participate in the rut for several consecutive years. Figure 4.10 shows Congal's ability to group hinds together over the 3 years 1974–1976, in comparison with the ability of another stag called Yesterday. The measure used in this comparison is hind-days, which is the number of hinds in the group multiplied by the number of days the group is kept together by the stag. If a stag has a group of two hinds for 1 day he has a hind-day score of two; if he has the group for 5 days, his score is ten; if he has a group of six hinds for 30 days his score is 180.

□ On the basis of the data presented in Figure 4.10, do you think that number of hind-days is a good measure of a stag's relative fitness?

■ It is not a very good measure. In the figure, Congal has a score of over 120 hind-days in 1974, but, as was mentioned above, he lost all his hinds in early October of that year, before mating began. His hind-day score was built up before his injury and is not a reliable indicator of his reproductive success for that year.

Despite the fact that these data do not provide a very accurate picture of Congal's reproductive success nor, for the same reasons, that of Yesterday, they do reveal something about the long-term effects of Congal's injury.

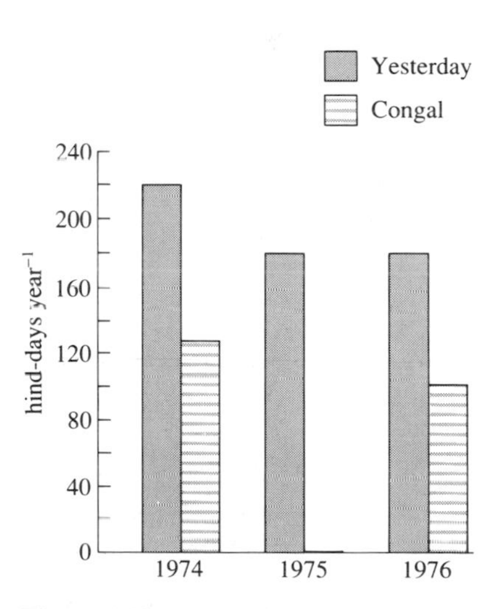

Figure 4.10 Reproductive success (measured in hind-days per year) for two stags, Congal and Yesterday, for the 3 years 1974–1976.

□ What do you infer from Figure 4.10 about Congal's reproductive success following his injury?

■ He grouped no hinds together at all in 1975 and so could have sired no calves in that year. However, in 1976 he had recovered at least some of his ability to group hinds and may, therefore, have produced offspring that year. It appears to have taken him two years to recover some degree of competitive ability.

This example illustrates very clearly one potential cost of fighting. An injury sustained in one rut totally eliminated Congal's reproductive capacity for 2 years.

□ Can you think of a reason why the data for Yesterday are included in Figure 4.10?

■ 1975 may have been a very bad year for all stags, not just Congal. To demonstrate that Congal's lack of success in 1975 was due to his injury and not to some general catastrophe, it is necessary to provide some evidence that other stags were successful in 1975. Yesterday was successful in 1975, at least in terms of hind-days, and so it can be concluded that Congal's failure was probably due to his injury.

This example illustrates that an injury incurred during fighting can lead to an animal leaving fewer offspring than it might have done. In other words, fighting can reduce fitness. Consequently, any genetic disposition to engage in disputes in a way which reduces the incidence or severity of injuries would be favoured by natural selection. This brings us back to the idea that behaviour performed during disputes might serve the function of enabling animals to assess one another's strength and so avoid a potentially dangerous fight.

Consider for a moment how adaptive it would be if animals could predict the probable outcome of a fight. Under such circumstances they would be able to enter only those fights which they predicted they could win, thus reducing their risk of injury from unnecessary, lost fights. We cannot assume that animals make *conscious* predictions, but they may respond to certain stimuli in such a way that dangerous fights are shortened or avoided.

4.6.2 The benefits of aggression

Benefits have to be measured in the same units as costs: fitness. Clearly, access to a resource, achieved through aggression, will tend to increase an animal's fitness and so constitutes a benefit. The greater the increase in fitness resulting from a dispute, the greater the benefit. Aggression is centred on some identifiable resource, such as an item of food, and the benefit to be derived from that aggression is obvious; the winner acquires the resource.

To illustrate the benefits of aggression in terms of individual fitness, the best examples are provided by those animals in which aggressive behaviour leads to an individual establishing possession of a territory. Figure 4.11 shows one such example, the sage grouse (*Centrocercus urophasianus*) of North America (Whiley, 1973). Sage grouse are a lek-breeding species (Book 1, Sections 4.2.1 and 9.3.3). Within a lek, males compete fiercely for the possession of territories, and fighting is fiercest for those territories nearest the centre of the lek. In Figure 4.11 those males holding central territories are identified by the lower numbers.

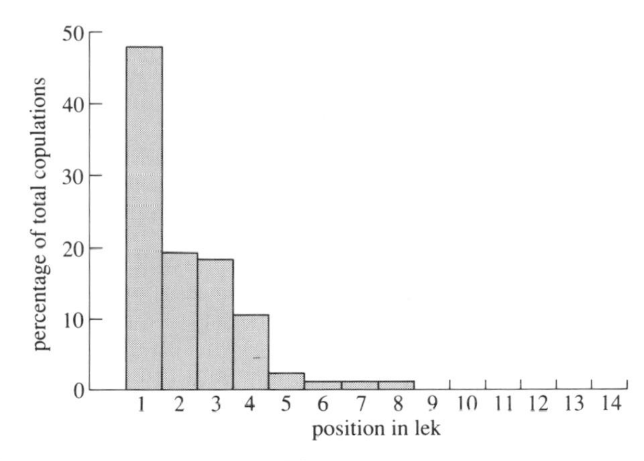

Figure 4.11 Reproductive success of fourteen male sage grouse in a single lek. Males identified by lower numbers occupied the more central territories in the lek. Of the 105 copulations that occurred in the lek, male 1 performed 48% of them.

- □ Do the data shown in Figure 4.11 provide evidence of the benefit of successful aggression? Explain your answer in terms of increases in an individual's fitness.

- ■ Yes, successful aggression results in males holding central territories. Females choose to mate with central males, not with peripheral males. Central males have the potential to sire more offspring than peripheral males, and thus have higher fitness.

In some species, the benefits derived from aggressive interactions may relate to long-term effects rather than to the immediate outcome of individual interactions. In those animals that live in stable social groups, individuals may learn to recognize one another and, on the basis of previous aggressive behaviour, a dominance hierarchy may be established (Book 1, Section 10.3). In a dispute over a resource such as food or a mate, an animal low in a dominance hierarchy will defer to one of higher rank. The benefits accruing to animals of high rank are obvious: they tend to gain priority of access to any resources that are in short supply without having to fight for them. In a study of bonnet macaques, the three highest-ranking males were responsible for nearly three-quarters of all copulations with receptive females. The high-ranking males formed exclusive associations and copulated with females that were also high-ranking. The low-ranking males, in contrast, did not become exclusive companions and rarely copulated with females at a time when they were likely to conceive.

- □ Can you think of a reason why it is of benefit to animals of *low* status to defer to those of higher status?

- ■ By doing so they avoid a fight which, on the basis of their prior experience, they would be very likely to lose.

It is important to stress that dominance hierarchies are not inflexible but change over time. As a general rule, younger animals tend to have low status and tend to rise in a hierarchy as they grow and become stronger. Likewise, ageing animals become weaker and decline in status. Thus, it is to the benefit of low-ranking animals to defer to high-ranking animals because they will avoid injuries which might reduce their future chances of rising in status.

Summary of Section 4.6

In principle, the consequences of aggressive behaviour can be assessed in terms of costs and benefits. Ideally, these would be measured in units of fitness but, given the constraints on observing animals in nature, they are more usually measured in terms of components of fitness. Studies of red deer have revealed some of the costs of injury incurred during fighting. Territorial species in which animals hold resources provide examples where it is relatively easy to measure the benefits of aggression. In species in which individuals form long-term dominant/subordinate relationships, benefits must be assessed in the long-term and not in terms of the immediate outcome of isolated disputes. It should not be assumed that animals can measure the costs and benefits of aggressive interactions directly.

Costs and benefits should be expressed in terms of decreases and increases in individual fitness, not simply in terms of resources lost or gained. This is because the benefit value of any given resource can vary according to circumstances. For example, in the case of the red deer, the value of 1 hind-day changes according to the stage of the rut. The benefit to the stag, in terms of increased fitness, of holding a hind for 1 day when she is sexually receptive is vastly greater than that gained by holding her for 1 day when she is unreceptive. However, it must be remembered that what the male does is only part of the story: the female too is a player in the mating game (Book 1, Chapter 9). For instance, she can elect not to mate with the grouse holding the central territory, or can escape from the stag attempting to keep her in a group.

The argument that is being proposed is that animals should evolve aggressive behaviour that maximizes the difference between the benefits and the costs of fighting. This is *not* intended to imply that animals can consciously assess costs and benefits in the same way that humans are able to. Rather, those individuals that, in the past, happened to behave in such a way as to maximize the difference between benefits and costs achieved the greater relative fitness. This raises a very interesting question. If it is found that animals do indeed behave *as if* they were good cost benefit analysts, how do they do it? A possibility which we have already raised is that certain components of aggressive behaviour may provide information by which animals are able to assess the relative strength of their opponent.

4.7 A modelling approach to animal aggression

If it were possible to measure accurately the costs and benefits of aggression, then it should also be possible to predict the most adaptive form of behaviour that an animal should adopt in an aggressive encounter. The net gain is defined as the benefit minus the cost, and is termed the **pay-off**. For example, if a small ram meets a large ram at a water hole at which there is room for only one animal to drink, it could be predicted that the costs to the small ram of disputing access to the water hole would be so great that it should not normally engage in a fight with the large ram. The only situation in which the small ram might be expected to fight would be if its life depended on its drinking immediately. J. Maynard Smith and G. Price (1973) in the UK developed the idea that optimum aggressive behaviour can be predicted, using a form of mathematics called *game theory*.

A game theory model enables one to predict which of a number of possible forms of aggressive behaviour, referred to as strategies, will yield the highest pay-off. Animals that adopt that strategy will be more fit than those that adopt other strategies, and so will be favoured by natural selection. Like all models in biology, game theory models represent a very simplified view of nature. Formulating and developing models can be a highly stimulating activity in itself, and models have many uses. One of those uses, particularly in research, is to generate predictions that can be tested by observations in natural situations. In the rest of this section the bases of a game theory model of aggression are introduced. How the model clarifies the nature of animal aggression is the subject matter of the subsequent section. (*Note* In the ethological literature, the model developed here is normally referred to as the *hawk–dove model*. These terms have been replaced here by fighter and threatener, to reflect more accurately the nature of the behaviour patterns of interest.)

As it is desirable to keep things as simple as possible, and because real animals are very rarely simple, the account begins by considering a population of hypothetical animals called snarks. A snark is like a robot automaton; its aggressive behaviour is rigidly stereotyped and completely under genetic control. Snarks have evolved aggressive behaviour that consists only of threat display; there is no transition from threat to fighting, or **escalation**, during an aggressive interaction. In a dispute between two snarks, the one that persists for longest with its threat is the winner.

Suppose that in a particular generation, a mutant snark appears in the population. It is a different kind of automaton and, when it engages in a dispute with another snark, it threatens briefly, then escalates and attacks its opponent. This mutant snark is called a fighter. If a fighter meets a normal snark, the normal snark always retreats when attacked. Hereafter, the normal snarks are referred to as threateners.

□ Will the mutant fighter snark win or lose disputes with other snarks?

■ It will win *all* its disputes, because all snarks it meets will be threateners, who will retreat when attacked.

Thus, the mutant fighter will gain access to all resources it disputes with other snarks. As a result, it will have very high fitness relative to threatener snarks, i.e. it will leave relatively more offspring.

□ How will the high fitness of the fighter mutant affect the abundance of fighters in subsequent generations?

■ The relative abundance of fighters will tend to increase in successive generations at the expense of threateners.

As fighters become more common, they will become increasingly likely to engage in aggressive interactions with other fighters, as well as with threateners. In a dispute between two fighter snarks, fighting continues until one is dead or so incapacitated that it can no longer fight. Consider the costs and benefits incurred by each kind of snark in the three kinds of dispute over resources that are now possible:

threatener against threatener,

threatener against fighter, and

fighter against fighter.

When two threateners meet, one of them gains the benefit represented by the disputed resource, both incur the same cost in terms of time spent in threat display, but neither incurs any cost in terms of injury. When a threatener meets a fighter, the threatener derives no benefit because it always withdraws, but it incurs negligible cost because the interaction is brief and it suffers no injury. The fighter gains the benefit of possession of the resource and incurs no costs.

□ In a dispute between two fighters, what are (a) the costs, and (b) the benefits, that accrue for the winner?

■ (a) High risk of injury and loss of time, and (b) possession of the resource.

□ In a dispute between two fighters, what are (a) the costs, and (b) the benefits, that accrue for the loser?

■ (a) High risk of injury and loss of time, and (b) none.

Thus, in a population in which there are substantial numbers of both fighter and threatener snarks, fighters experience a high benefit: cost ratio whenever they meet a threatener, but a low benefit: cost ratio when they meet another fighter. As the frequency of fighters in the population increases from generation to generation, high cost fights between fighters become more common. Thus, an individual fighter will have lower relative fitness in a population in which fighters are common than it will in a population in which they are rare. As a result, natural selection will favour fighter snarks less strongly, the more common they become. Over several generations, the relative abundance of fighters in the population will stabilize at a certain ratio of threatener to fighter snarks. The precise value of this ratio will depend on the actual values of the benefits and costs in terms of fitness. If the possession of disputed resources can greatly enhance fitness, and if fighting costs are low, stability will be reached at a point at which fighters are very abundant. Conversely, if resources contribute little to fitness and fighting costs are high, fighters will always be rare.

An important point emerges from this simple model—the cost:benefit ratio of any one strategy depends on what other strategies are being employed in the population, and on how common they are. The fitness of a fighter snark does not have an absolute value, but depends on how common fighters are, relative to threateners.

It is possible to perform mathematical calculations based upon the kind of assumptions just discussed. The details go beyond this course but the principles that emerge are highly relevant. It turns out that at one particular ratio of threatener to fighter numbers, the relative fitness of fighter and threatener is equal.

□ What does having an equal relative fitness mean in terms of the ratio of fighters to threateners in the population over several generations?

■ It means that threateners and fighters will leave the same number of offspring in the next generation. Thus, the ratio at which threatener and fighter have equal relative fitness is stable.

Three important points emerge from developing such a model:

1 As noted earlier, the pay-off to an individual of one type depends on how common the other type is in the population.

2 The ratio at which the population stabilizes depends on the values of the pay-offs, which, in turn, depend on the values of the costs and benefits of fights. With different values for costs and benefits the stable ratio would be different.

3 The example that was described envisaged a situation in which individual snarks belonged to one of two distinct types—fighter and threatener. A different 'game' could have been developed in which all snarks are the same and behave as threateners in some of their fights and as fighters in others. Such snarks would be more like real animals. In this case the model can be used to determine how an individual should apportion its time between being a threatener and fighter. This kind of variable strategy is more realistic because, in real animals, individuals are more likely to escalate a fight at certain times than at others.

Suppose that a population of snarks *has* evolved in which there are certain costs and benefits attached to fighting and threatening. Suppose also that each individual devotes time to behaving as a fighter and as a threatener in a certain ratio, for example 60% of the time as a fighter and 40% as a threatener. Suppose also that, with specified benefits and costs, calculations show that any departure, in either direction, from this ratio yields lower pay offs. In other words, individuals that fight more or less than the rest of the population are less fit. The majority strategy is called an evolutionarily stable strategy (ESS). An ESS is defined as a pattern of behaviour compared with which no alternative strategy can yield a higher fitness. In this example, animals adopting an exclusively threatener or exclusively fighter behaviour will be less fit than those adopting fighter and threatener behaviour in the ratio of 60%:40%. The ESS in this model is a mixed strategy—the animals with the greatest fitness are not exclusively threatener or fighter, but rather use both strategies at different times.

What are the relative values of costs and benefits in nature? Costs of injury could be relatively very low in animals that lack weapons of any kind. Male toads, who fight over females, have no teeth, claws or other weapons, but indulge in struggles that may last many hours. For most animals, it is very unlikely that the fitness gained as a result of fighting will exceed the potential loss of fitness due to injury, simply because animals generally have other opportunities to fight for the resource that will provide the fitness gain. Only if all, or a large proportion of an individual's fitness depends on the outcome of a single fight, is such a situation likely. Thus if the benefits are high enough, animals would be expected to incur high costs to obtain them. In fig wasps (*Idarnes* sp.), males have but one chance to mate in their brief lives. Fighting between males over females is violent and often lethal. Those males that are successful reproduce, those that are not successful do not.

Summary of Section 4.7

Game theory provides a theoretical model by which the fitnesses of different aggressive strategies can be compared. In a two-strategy model, the fitness for an individual adopting one strategy depends on the abundance of individuals adopting the alternative strategy. When costs due to injury exceed the benefits of fighting, a strategy that incorporates a tendency to escalate a dispute towards fighting has higher fitness than one involving more limited aggression, but only so long as the proportion of animals in the population showing the strategy is limited. A model in which a population consists of two distinct types, each having its own stereotyped aggressive strategy, is mathematically equivalent to a more realistic situation in which all individuals can adopt both strategies. Game theory predicts how much time an individual should devote to each strategy. The behaviour shown by individuals of fighting some of the time and threatening some of the time in a particular ratio which gives the highest possible fitness to the individual is an ESS.

4.8 Asymmetric contests

In developing the game theory model, it was assumed that animals differ only in the strategy (e.g. fighter or threatener) that they show during aggressive interactions. The model took no account of the possibility that two animals involved in a dispute might differ in other respects, e.g. size, age, etc. In the mid-1970s, Geoffrey Parker of the University of Liverpool pointed out that such symmetric contests visualized in simple game theory models will generally be very rare in the real world. A much more common situation will be one in which one animal has some intrinsic advantage over the other. It might be bigger, more experienced, less tired, or simply better fed. An aggressive interaction in which there is some disparity of this kind is called an **asymmetric contest**.

Asymmetries between animals can be incorporated into game theory models, which become mathematically more complex as a result. We shall not go into the mathematics but will concentrate on the kinds of asymmetries that have been considered, the predictions that game theory makes about them, and examples of aggression in real animals that fit those predictions.

4.8.1 Correlated asymmetries

If two animals differ in their ability to win a fight because, for instance, one is bigger than the other, this is called a **correlated asymmetry**. In other words, the asymmetry, the difference between the animals, is an intrinsic property of the animals themselves. (You will see the significance of the term *correlated* asymmetry when *uncorrelated* asymmetries are introduced in the next section.) One kind of asymmetry that may exist between two animals is the ability to gain and to maintain possession of essential resources, referred to as **resource holding potential**.

Resource holding potential asymmetries arise when animals differ in weight, size, weapon size, experience, or even position on the local terrain. An animal that has established a position on top of a hillock may be in a better position to win a fight than one who has to attack it from below. Game theory makes a simple prediction about contests in which there is resource holding potential asymmetry. The animal with the higher resource holding potential is, by definition, more likely to win a fight and is less likely to be injured. In other words, it is more likely to accrue a fitness gain and less likely to incur a cost. Therefore, we can predict that an animal with higher resource holding potential than its opponent should adopt a strategy involving a high propensity to escalate. Let us now consider a real example.

European common toads (*Bufo bufo*) spend most of the year on land, but each spring they migrate to ponds to breed. Males, who are more abundant than females in a breeding pond, clasp onto the backs of females and are carried about in this position (called amplexus) for several days. Eventually, the female produces her eggs in two strings, entwining them around vegetation in the pond. As the eggs emerge, a process called spawning, the male fertilizes them. During the 3 or 4 days that a male is in amplexus, he is at risk of being displaced by another, unpaired, male as a result of a fight which takes place on the female's back. Such a displacement is called a take-over. Nick Davies and Tim Halliday (1979) working in Oxford recorded which males achieve take-overs.

They classified male toads into five size classes, and found that all males had an equal chance of being the first to clasp a female, irrespective of their size. However, by the time that females started to spawn, large males were much more likely to be in amplexus than small ones (Figure 4.12).

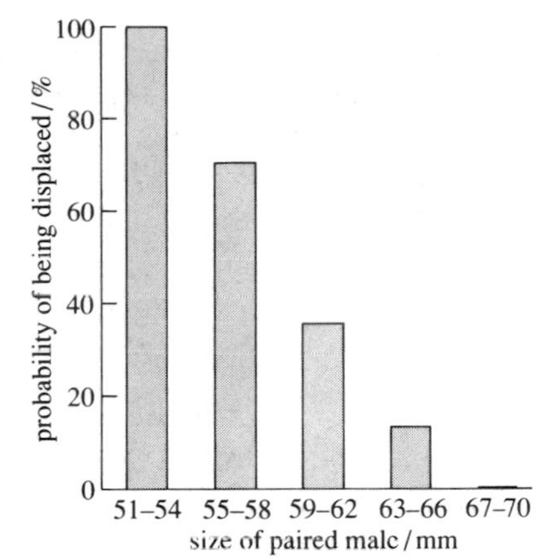

Figure 4.12 The probability with which male toads of different sizes (measured from the snout to the vent, the vent being the excretory/reproductive opening) will be displaced from the backs of females by other males between the time that they first clasp the female and the time that she spawns.

□ What is the relationship between size and the resource holding potential of male toads?

■ The larger a male is, the greater his resource holding potential.

Consider a medium-sized male toad who is not holding a female.

□ Should he show a higher propensity to attack and attempt to take over the female held by a male larger or a smaller than himself?

■ A smaller male. It has lower resource holding potential than him, whereas the larger male has higher resource holding potential.

Davies and Halliday set up experimental fights to test this prediction. Male toads were classified as being large, medium or small. Twelve males of each size class were allowed to clasp females; these paired males were the defenders in fights against attackers, of whom there were 12 in each size class. Thus there were nine kinds of fight (large defender vs large attacker, large defender vs medium attacker, etc.) and 12 of each kind were observed. During observation of each fight, the behaviour of the attacker was recorded and a measure of his persistence in attacking was obtained. The results are shown in Figure 4.13.

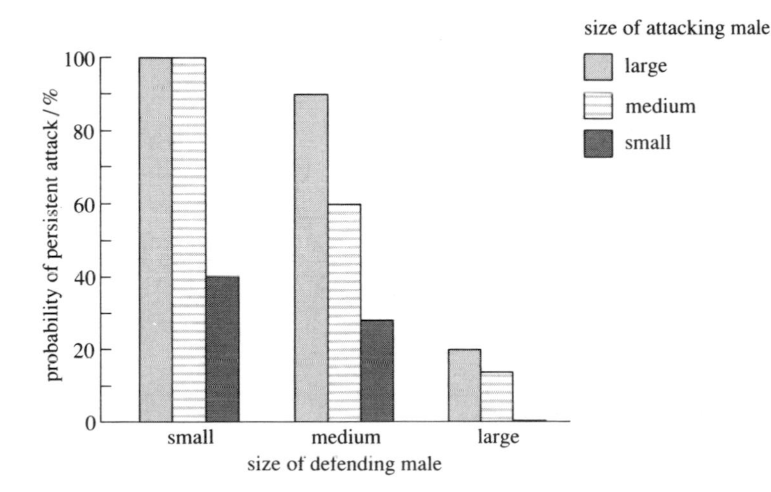

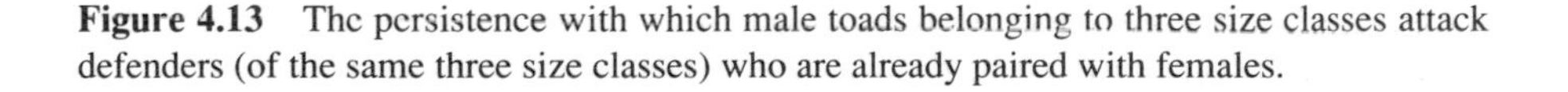
Figure 4.13 The persistence with which male toads belonging to three size classes attack defenders (of the same three size classes) who are already paired with females.

☐ Do these results support or refute the prediction that an attacker's tendency to fight should be related to the difference in resource holding potential between him and the defender?

■ The prediction is supported. Attackers of all three size classes were more persistent against defenders smaller than themselves than they were against larger defenders.

Male toads thus behave in the way that game theory predicts. Their attacks are strongest against those opponents that have lower resource holding potential than themselves.

4.8.2 Asymmetry and assessment

Game theory predicts that an animal's tendency to attack an opponent should depend on the relative resource holding potentials of the attacker and the opponent, and male toads fulfil this prediction. In order to do this, animals need to be able to assess the resource holding potential of their opponents. The concept of assessment was raised earlier in this chapter, and evidence that it actually occurs in animals is examined below.

Clearly, an animal can gain an assessment of an opponent's resource holding potential by fighting with it; if it gets beaten it obviously has lower resource holding potential. However, it would be adaptive if it could make its assessment before any fighting begins, because it could then avoid a fight and thus not incur any of its attendant costs. This argument leads to a prediction about aggressive behaviour. If animals perform behaviour patterns early in aggressive interactions that provide reliable indicators of their resource holding potential, such behaviour patterns should be used by their opponents to determine that opponents' propensity to attack them. Examples of animals in which this prediction has been tested are red deer and toads.

During the rut, an aggressive interaction between two red deer stags commonly begins with them standing some distance apart, each roaring several times in succession and in turn, first one stag, then the other. Tim Clutton-Brock and others have investigated these 'roaring contests' and found that stags roar at each other with a gradually increasing tempo. When one stag roars at a much higher rate than the other, no fight ensues, and the two stags move apart. The stag which roared at the higher rate retains possession of his hinds or territory. This roaring is interpreted as a form of assessment. Since it probably requires considerable strength to roar at a great rate, a stag which can roar at a high rate is likely to be a strong stag. Evidence that this is so comes from the observation that stags generally roar at a higher rate early in the rut when they are in good condition than they do towards the end when they have become exhausted by fighting and lack of nourishment.

The hypothesis that stags use roaring to assess one another's strength was tested by a field experiment in which recordings of roars were played to stags through loudspeakers. The recordings were arranged in such a way that the taped stag roared at a higher rate than the stag to whom the recording was being played. Faced with this stimulus, most stags stopped roaring after a while and retreated, suggesting that indeed they responded to the roaring of an apparently more powerful rival.

In toads, as you have seen, the ability to win a fight is dependent on a male's size relative to that of his opponent. Is it possible that the reason why male toads are less persistent in attacking males larger than themselves is that they are able to assess the size of their opponents? One obvious way that they might do this is by looking at their opponent. Also, when toads fight they kick out at each other with their hind legs. Larger toads probably deliver more powerful kicks than small ones. Finally, while fighting, male toads continually emit soft, peeping croaks. Davies and Halliday (1978) proposed and tested the hypothesis that these croaks are used in size assessment.

They first recorded and analysed the calls made during fighting by males of different sizes, and found a strong relationship between size and call frequency (pitch). As shown in Figure 4.14, large males make deep croaks, small males make high-pitched ones. Thus a male's call *could* convey reliable information to a rival about his size. The question is, does the rival actually use this information?

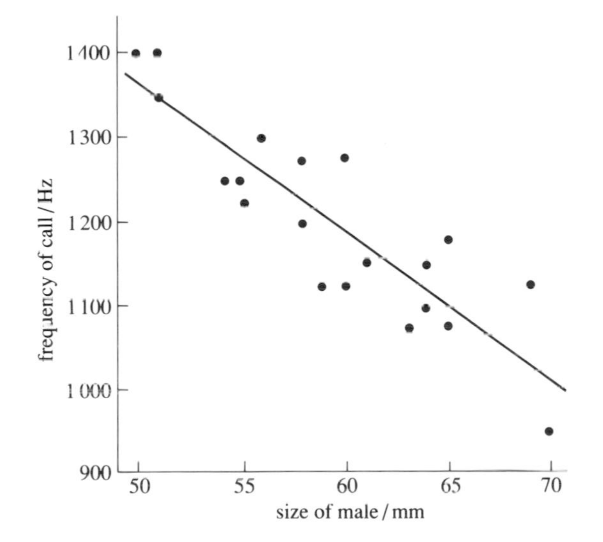

Figure 4.14 The relationship between male size (measured from the snout to the vent) and the frequency of the call made during fighting in a sample of 20 male toads.

Twenty-four medium-sized males were collected from a pond. Twelve of these were used as attackers against small males paired with females (defenders), with the other 12 against large defenders. The defenders were temporarily silenced by fitting them with rubber bands, placed behind their arms and passing through their mouths like a horse's bit. This prevented them from croaking but did not affect their behaviour in any other way, nor did it harm the males. All of the females were of similar size. Each attacker was used in two trials against the same silenced defender. Trials were conducted in a small tank containing the pair, the attacker and a small loudspeaker positioned just above the pair. Recorded croaks, of either small or of large males, were played through the speaker for 5 seconds whenever the defender was touched by the attacker. Since defending males were either small or large, the call that was played from above them was either appropriate or quite inappropriate to their actual size. The behaviour of attackers was recorded to see whether the real or the apparent size (as indicated by the recorded call) influenced attackers' behaviour. The results are shown in Figure 4.15 (*overleaf*).

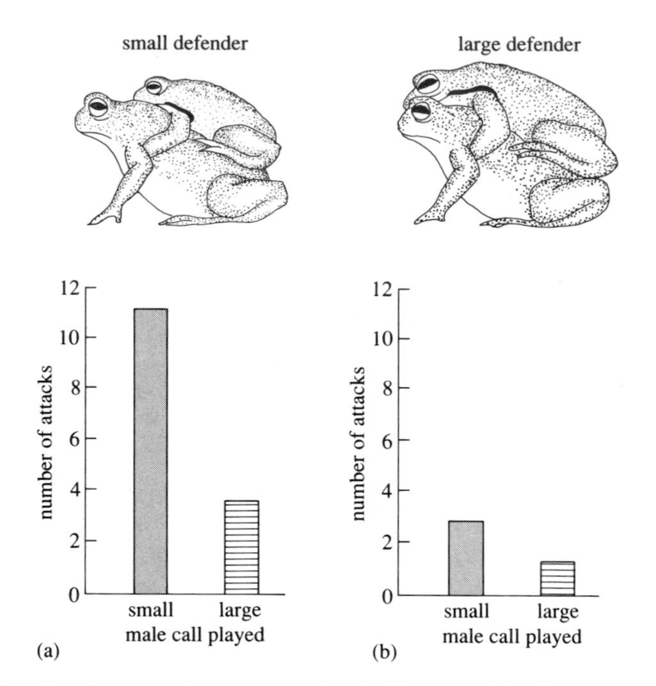

Figure 4.15 Results of an experiment to test the significance of the frequency of the call produced during fighting by male toads, (a) showing a small defender, and (b) a large defender. In each pair of histograms, the shaded column summarizes results of trials in which the call of a small male was played, the hatched column those in which the call of a large male was played.

Consider the histograms in Figure 4.15a, which shows the number of attacks given against small defenders.

□ Which type of call, that from a small or that from a large male, elicited the *higher* level of attack?

■ Calls from small males.

Now consider the histograms in Figure 4.15b, which show the number of attacks against large defenders.

□ Which type of call, that from a small or that from a large male, elicited the *lower* level of attack?

■ Calls from large males.

These results provide strong evidence that the nature of the call heard by an attacker influences the rate at which he delivers his attacks. However, if the call was the *only* cue that influenced an attacker's behaviour, the prediction would be that when comparing fights in which the calls were the same, small and large defenders would be attacked with equal frequency. In other words, the *actual* size of the defender would not influence an attacker's behaviour. Compare the two pairs of histograms in Figure 4.15a and b.

□ Is this prediction supported by these data?

■ No. Comparing trials in which the recorded calls were the same, there were fewer attacks directed towards the large than the small defenders.

Clearly, the call is not the *only* cue that influences an attacker's behaviour. Whereas the playing of croaks of a large male inhibited attacks against a small defender, playing croaks from a small male did not elicit as many attacks on large defenders as it did on small defenders. Croaks are, therefore, used in assessment, but are not the only cue used. Male toads probably also respond to the actual size and the strength of kick of their opponents.

4.8.3 Uncorrelated asymmetries

An **uncorrelated asymmetry** is a disparity between two animals that is not due to their intrinsic properties but is due to chance or some external influence. If two people settle possession of the last biscuit on a plate on the basis of which of them is the larger, they are using a correlated asymmetry. If they settle on the toss of a coin, they are using an uncorrelated asymmetry; there is no relevant difference between them other than that one of them called correctly.

Consider a situation in which two animals approach a resource, such as an item of food. They are reasonably well matched for size and strength, and the benefit to be gained from the resource is small. They could settle their dispute by means of threat or some other form of behaviour that enables each to assess the relative strength of its opponent. However, if the two animals are of equal or very similar strength, threat may escalate to a fight involving possible injury; if they differ in strength they may have spent considerable amounts of time and energy in making their assessments. Whatever happens, they may both incur costs that exceed the benefit value of the resource. In this kind of situation, game theory predicts that if the benefit is low and the potential cost is high, the ESS is for the two animals to settle the dispute quickly by means of some uncorrelated asymmetry. In the human analogy, it is obviously more sensible to settle the dispute over the last biscuit by tossing a coin than it is to fight over it.

Animals do not toss coins, but a simple uncorrelated asymmetry they could often use is which of them gets to a resource first. This is much more likely to be a matter of chance than it is to be a reflection of any superiority that one animal has over another. An example in which priority of possession is used by animals in this way is the speckled wood butterfly (*Pararge aegeria*).

Speckled wood butterflies were studied by Nick Davies (1978) in a wood near Oxford. Males claim and defend sunlit patches on the floor of the wood, where they court females who are attracted to such 'sunspots'. There are never enough sunspots for all the males to occupy at any one time and there is always a number of males moving about in the leafy canopy of the wood. When one of these sees a vacant sunspot he flies down and occupies it. Sunspots are not stable resources; as the sun moves across the sky, existing ones move, grow, contract and disappear and new ones appear. From time to time a male from the canopy flies down into an occupied sunspot, where he is challenged by the owner. The two males make a brief spiral flight up towards the canopy, after which one continues up into the canopy and the other flies back down and settles in the sunspot. By marking individual male butterflies, Davies was able to show that it is invariably the original owner who returns to a sunspot after a spiral flight, even though the owner may only have been there for a few minutes.

There are at least two possible explanations of such behaviour. The first is that a spiral flight somehow serves to inform an intruder that a sunspot is occupied. The second is that only relatively strong butterflies hold sunspots and that a spiral flight serves to demonstrate their strength. Davies favours the first explanation, that an intruder somehow recognizes and respects prior ownership. One reason for doubting that only strong butterflies hold sunspots was that most of the males he marked when they were in the canopy were subsequently observed holding sunspots. To test his hypothesis, he performed an experiment in which he removed the owner from a spot, waited until a new male descended from the canopy to occupy it, and then released the original owner back into the spot. On each occasion, there was a brief spiral flight, after which the original owner flew up to the canopy and the new owner returned to the sunspot (Figure 4.16). Were the possession of a sunspot dependent on male strength, the original owner would surely have been able to reclaim it from the new owner, in at least some of the trials. The result of this experiment suggests that male speckled wood butterflies settle territorial disputes on the basis of the uncorrelated asymmetry of prior ownership, and that a male need only occupy a territory for a few seconds to become the owner.

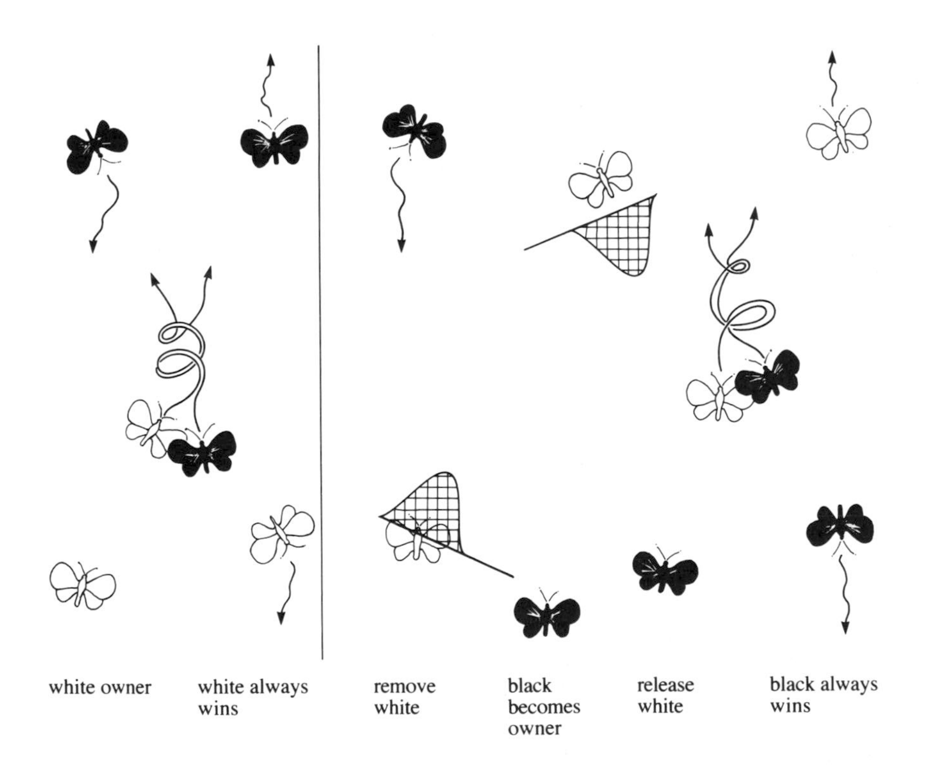

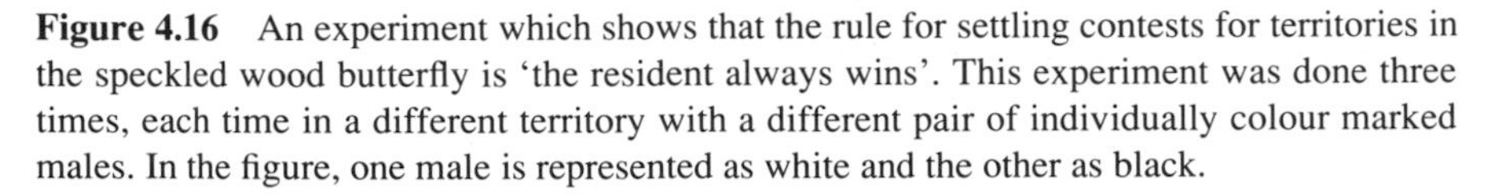

Figure 4.16 An experiment which shows that the rule for settling contests for territories in the speckled wood butterfly is 'the resident always wins'. This experiment was done three times, each time in a different territory with a different pair of individually colour marked males. In the figure, one male is represented as white and the other as black.

This behaviour can be interpreted in terms of costs and benefits. Any kind of prolonged contest between butterflies is likely to incur high costs.

□ What might these costs be?

■ There are three main costs.

1 There is the energetic cost, the energy used in the contest. This cost also results in time being spent away from sunspots replacing the energy by feeding in the canopy.

2 There is the cost of absence. A prolonged contest takes the contestants away from the forest floor and may mean a prospective mate being missed, or a third male taking over the sunspot.

3 There is a physical cost. Close interaction between two males may lead to damage to their delicate wings or flying muscles.

Benefits are likely to be low because sunspots are highly ephemeral; a male who takes over a sunspot may very well find that it soon shrinks and disappears. Furthermore, males who are not holding a sunspot at any one time have only to wait for new ones to appear. As was argued earlier, it is in situations in which potential benefits are much smaller than potential costs that disputes would be expected to be settled by means of uncorrelated asymmetries.

If these butterflies do somehow recognize and respect territory ownership, what happens if two butterflies are experimentally confused so that each considers itself to be the owner? The prediction would be that, as neither recognizes the other as the owner, neither should give way and retreat to the canopy. Davies investigated this by surreptitiously introducing a second male into an already occupied sunspot. This can be done in larger sunspots. The intruder, undetected by the real owner, settled in the sunspot. Sooner or later one of the males noticed the other and challenged it. In all cases there ensued a protracted spiral flight, lasting on average ten times longer than normal flights. It appears that a male that perceives itself to be the owner is prepared to continue with the spiral flight until the other gives up. When only one male perceives itself to be the owner, only he continues and the flight is short; when both claim ownership, both continue and the flight is long.

Summary of Section 4.8

The section began by considering the fact that two animals will rarely be perfectly matched for fighting ability or resource-holding potential. This led to the prediction that such correlated asymmetries should be used as a basis for settling fights quickly, and thereby reducing the costs to the protagonists. Studies of male toads support this prediction; individual males are more persistent when attacking those opponents that they are most likely to defeat by virtue of their relative size. Game theory also predicts that correlated asymmetries should be used by animals as a means of assessing their opponents. Experiments conducted on the vocalizations of red deer stags and male toads support this prediction. Finally, game theory predicts that, under certain circumstances, uncorrelated asymmetries may be used to settle aggressive interactions quickly. This prediction is supported by experiments on male speckled wood butterflies, in which prior ownership of a sunspot is the asymmetry.

Overview of Sections 4.5–4.8

Taken at face value, such commonly heard expressions as 'the survival of the fittest' and 'the struggle for existence' suggest that animals would fight to the limits of their capabilities. However, observation of aggressive interactions between animals reveals that in most species aggressive behaviour leads to the settling of disputes by means of displays rather than actual fighting. When fighting does occur, there is a real possibility of injury. Sections 4.5–4.8 have examined one particular approach towards answering the question of why species vary in the form of the aggressive behaviour that they perform.

The basis of the argument is that one can, in principle, identify and measure costs and benefits resulting from different kinds of aggressive interaction. In some species this exercise has proved to be possible. Game theory can be used to predict, on the basis of known costs and benefits, the most adaptive form of aggressive behaviour that a species can employ. Finally, experiments with real animals have borne out the predictions of game theory.

Sections 4.5–4.8 have been concerned with the evolution of aggressive behaviour. They have considered the basic question of what form of aggression is most adaptive in a given situation. From the argument that natural selection will favour a particular form of aggression in a certain species, you should not draw the conclusion that that form of behaviour is necessarily determined genetically. In fact, as we have seen, aggressive behaviour *is* variable. This does not invalidate the game theory approach, which simply seeks to predict the most adaptive form of aggressive behaviour in a given set of conditions. Whatever the processes that determine the development of aggression during an animal's life, it should still behave in the way that game theory predicts. For example, animals within a dominance hierarchy behave in a conventional way, subordinates giving way to those of higher rank (Section 4.6.2). Such behaviour is learned as a result of experience gained in previous aggressive interactions. Furthermore, animals alter their behaviour as their position in a dominance hierarchy changes. Thus, although their behaviour is not rigidly stereotyped but changes during their life, they still conform to the predictions of game theory.

Aggression was defined as behaviour that enables individuals to acquire or gain access to resources at the expense of others. The general conclusion to draw from the arguments and data presented in Sections 4.5–4.8 is that, when the value of disputed resources is high, animals will tend to show more dangerous forms of aggressive behaviour than they will when resources have low value.

4.9 Relevance of game theory to human aggression

The arguments presented here for the function of animal aggression can be applied to certain aspects of human aggression (e.g. some contests between individuals) in western society, but only after three major modifications. Firstly, competition in western society is not generally about competition for essential scarce resources. People do compete, but for different resources. For instance, when in a hurry they compete for time: they also compete for jobs, and they compete when trying to persuade you to buy their product rather than someone else's. In everyday language such competition is often called aggressive, e.g. an aggressive driver, an aggressive salesman. Secondly, western society is biologically perverse in that people do not maximize the number of offspring they can produce. Thus the bedrock of evolutionary theory, fitness, cannot be applied in any straightforward way to current human behaviour. However, if status and power are substituted for fitness, it is possible to refer to increases in status consequent upon competition. (It could be argued that maintenance of social standing is the basis for some brawls.) This substitution allows the third modification, namely that costs and benefits are measured in units of status and power, not in units of fitness. People are very good at assessing the costs and benefits of their behaviour, usually in terms of money or social standing or some other measure of partial status and power. (These arguments should not be extended to larger groupings of people, e.g. nations, as too many other factors are involved in these contests.)

Clearly, these modifications radically alter the model. However, the general statement of the model, that animals compete for resources and assess costs and benefits, is applicable to all animals, including people.

Summary of Chapter 4

This chapter emphasizes that aggression is under the control of factors internal and external to the animal. A principal internal factor (testosterone) was considered, followed by a brief look at where in the brain integration of external and internal factors might take place. Whilst the evidence for an organizing and motivational effect of testosterone seems quite consistent, the effect of testosterone on aggression in the adult male is dependant on numerous factors. The study of the neural substrates of aggression is becoming more refined as the neurophysiological and behavioural techniques available become more sophisticated. The problems of experimental studies of aggression and, in particular, the behavioural measures to use was a recurring theme which has still not been satisfactorily resolved, essentially because no one experimental method can reflect the complexity and variety of aggressive behaviour. The evolution of aggression was briefly considered, followed by the game theory model of function. It focused on the kinds of external stimuli an animal ought to be responsive to if it is to maximize benefits and minimize costs and so increase its fitness. Evidence in support of these predictions was presented showing that animals do appear to follow certain rules of engagement which increase fitness.

The most important message from this chapter is that, with a topic like aggression, it is essential that it is considered from a number of different perspectives. As it is, certain aspects have been omitted from this chapter, such as the neurochemistry and the social psychology of aggression. This does not mean that these areas are not important, but merely that space is limited. The topic of aggression is necessarily multidisciplinary because it needs to be studied at different levels, which means that there can never be simple answers to the questions of aggression raised in Section 4.1.2. Yet the simple answer is often sought.

Objectives for Chapter 4

When you have completed this chapter, you should be able to:

4.1 Distinguish and use, or recognize definitions and applications of each of the terms printed in **bold** in the text. (*Question 4.1*)

4.2 Distinguish between the various definitions of aggression. (*Questions 4.1 and 4.9*)

4.3 Present evidence for the effects of hormones on aggression, distinguishing between the organizing and motivational effects of hormones. (*Question 4.2*)

4.4 Discuss the issues pertaining to the study of neural substrates and aggression. (*Question 4.3*)

4.5 Discuss the role of environmental context as a factor in determining aggression. (*Question 4.4*)

4.6 Explain aggression in terms of its function. (*Question 4.5*)

4.7 Describe the assumptions underlying game theory. (*Questions 4.6 and 4.7*)

4.8 Present evidence that animals assess each other during aggressive interactions. (*Question 4.8*)

4.9 Separate functional, causal and definitional issues when considering human aggression. (*Question 4.9*)

Questions for Chapter 4

Question 4.1 (*Objectives 4.1 and 4.2*)
State the style of definition of aggression used in the following:

(a) A behaviour pattern is aggressive if it leads to an unequal division of resources.

(b) A behaviour pattern is aggressive if it involves the display of a structure that can inflict harm on an opponent.

Question 4.2 (*Objective 4.3*)
Give three reasons why the effect of the hormone testosterone on aggression may differ between adult males of the same species.

Question 4.3 (*Objective 4.4*)
In the study by Kruk *et al.* reported in Section 4.4.2, there were 217 electrode placements in the hypothalami, which had either no effect, or no predictable effect on the behaviour pattern being measured. How would you account for this result?

Question 4.4 (*Objective 4.5*)
In behavioural terms, why is it particularly difficult to relate studies of animal aggression to human aggression?

Question 4.5 (*Objective 4.6*)
State briefly why a pattern of behaviour in which an animal benefits its social group but disadvantages itself is said not to be an ESS.

Question 4.6 (*Objective 4.7*)
For each of the following statements, state whether it is true or false. If you think the statement is *false*, briefly state the reason.

(a) It is always adaptive for individuals to escalate in aggressive interactions.

(b) The pay-off from a fight is defined as the benefit derived from that fight.

(c) In a species in which there has evolved a form of fighting that represents an ESS, no further evolutionary change in fighting behaviour will occur.

(d) The question of which of a number of possible aggressive strategies represents an ESS can be resolved simply by working out the pay-offs accruing to each kind of individual in each possible kind of fight in which it might engage.

Question 4.7 (*Objective 4.7*)
Aggression between pairs of male jumping spiders consist of a number of stages of different intensity which can be ranked from 1 to 5. Would you expect contests between males to be more or less intense in the presence of females? Explain your answer in terms of costs and benefits.

Question 4.8 (*Objective 4.8*)
What evidence would you cite in support of an argument that animals assess the resource holding potential of a rival before fighting?

Question 4.9 (*Objectives 4.2 and 4.9*)
One definition of human aggression that is sometimes used (e.g. in the Open University course *Introduction to Psychology*) is in three parts:

The aggressor must have an intention to harm the victim.

The victim must be another living thing.

The victim must be motivated to avoid such treatment.

This is quite a good definition, but there are two reasons why it is not universally acceptable. What are they?

References

Adams, D. (1971) Defence and territorial behaviour dissociated by hypothalamic lesions in the rat, *Nature*, **232**, pp. 573–574.

Christensen, T., Wallen, K., Brown, B. and Glickman, S. (1973) Effect of castration, blindness and anosmia on social reactivity in the male Mongolian gerbil, *Physiology and Behaviour*, **10**, pp. 989–994.

Clutton-Brock, T. H., Albon, S. D., Gibson, R. M. and Guinness, F. E. (1979) The logical stag: adaptive aspects of fighting in red deer (*Cervus elephus* L.), *Animal Behaviour*, **27**, pp. 211–225

Davies, N. B. (1978) Territorial defence in the speckled wood butterfly (*Pararge aegeria*); the resident always wins, *Animal Behaviour*, **26**, pp. 138–147.

Davies, N. B. and Halliday, T. R. (1978) Deep croaks and fighting assessment in toads, *Bufo bufo*, *Nature*, **274**, pp. 683–685.

Davies, N. B. and Halliday, T. R. (1979) Competitive mate searching in male common toads, *Bufo bufo*, *Animal Behaviour*, **27**, pp. 1253–1267.

Kruk, M. R., van der Poel, A. M., Meelis, W., Hermans, J., Mostert, P. G., Mos, J. and Lohman, A. H. M. (1983) Discriminant analysis of the localization of aggression inducing electrode placements in the hypothalamus of male rats, *Brain Research*, **260**, pp. 61–79.

Lumia, A., Westervelt, M. and Rieder, C. (1975) Effects of olfactory bulb ablation and androgen on marking and agonistic behaviour in male Mongolian gerbils, *Journal of Comparative and Physiological Psychology*, **89**, pp. 1091–1099.

Maynard-Smith, J. and Price, G. (1973) The logic of animal conflict, *Nature*, **246**, pp. 15–18.

Whiley, R. H. (1973) Territoriality and non-random mating in the sage grouse *Centrocercus urophasianus*, *Animal Behaviour Monographs*, **6**, pp. 87–169.

Further reading

Archer, J. (1988) *The Behavioural Biology of Aggression*, Cambridge University Press.

Groebel, J. and Hinde, R. A. (1989) *Aggression and War: Their Biological and Social Bases*, Cambridge University Press.

Huntingford, F. A. and Turner, A. K. (1987) *Animal Conflict*, Chapman and Hall.

CHAPTER 5
STRESS

5.1 Introduction

This chapter continues the broad theme of Book 5: How is behaviour controlled by external and internal factors? The last chapter considered the reactions of animals when subjected to a threat of some kind. Aggression following a threat was discussed at length, but fear reactions were also mentioned briefly. In some respects the present chapter follows logically from Chapter 4, considering the response to threat but with a different emphasis. Whereas the situations described in the last chapter were ones in which an appropriate response could generally be made, Chapter 5 considers the way in which an animal reacts when external stimuli are such that no obvious behavioural solution is available. This situation is described here under the heading of 'stress'. Before looking at a model of stress deriving largely from animal experimentation, it is useful to relate the concept to everyday experience.

5.1.1 What is stress?

Surely most, if not all, people have described themselves as being under stress at some time in their lives. Indeed, the lot of many Open University students in simultaneously studying, raising a family and being in (or out of) employment doubtless brings its share of stress. Books and articles on stress abound. Numerous popular books available on stands at railway stations inform long-suffering commuters how to beat stress or how to live with it. The scientific literature contains over a million articles on the topic. However, a widely accepted definition of the word stress still eludes the scientific community. When is an individual stressed? Often scientific discussions of this issue get into deep water and are forced to fall back onto intuitive feelings based upon personal history. In clinical diagnosis, the personal experience of the 'stressed' person counts for much. Before you go on, you might like to stop for a moment and think of some situations that you have experienced that you regard as involving stress. Try to write down briefly what made them stressful.

Doubtless, you will arrive at a rich variety of different situations. Some representative ones that many would regard as stressful include: (a) being stuck in a traffic jam on the way to an important engagement, (b) having a deadline to meet and too little time, (c) being exposed to noise night after night so as to preclude sleep, (d) falling into debt, and (e) chronic relationship problems.

Stress is a concept central to human life, but it is also relevant to animals. Consideration of domestic animals soon reveals an abundance of situations which, on the basis of empathy and the normal behaviour of the animals, one would have little difficulty in describing as stressful. For example, (1) sows tethered for many weeks on short chains, with severely limited movement, (2) veal calves kept in small wooden boxes that preclude the animal turning around, (3) restrained animals that commonly engage in repetitive and destructive (to self and to others) behaviour patterns (in the Netherlands alone each year, 100 000 fattening pigs are lost as a result of infection following tail biting by penmates), and (4) animals destined for slaughter, transported across continents in cramped conditions.

In recent years, psychologists and applied ethologists have become closely involved in questions of animal welfare and suffering. Stress is often taken as synonymous with suffering and poor welfare. The EEC has sponsored a number of conferences at which criteria for defining stress were discussed, and recommendations for European Community legislation were aired. In the mid-1980s, Cambridge University appointed Britain's first chair of animal welfare with a research base in the area of stress. Sweden has pioneered new approaches to animal husbandry, closely guided by applied ethologists.

Although subjective impressions and empathy play a valid role, scientific criteria for determining when an animal is stressed are also needed. This is particularly so when arguing the case for appropriate legislation. Given that over a million articles have so far failed to deliver a universally acceptable definition of stress, you should neither fear that you will be asked to produce one nor even have high hopes that this chapter can deliver such a definition. However, some clear pointers can be given, and a broad consensus has emerged in behavioural science.

A number of different approaches to stress, and to the criteria used to decide when an animal is stressed are commonly given, with reasonable correspondence between them.

1 Investigators can look at the physiology of the body. Situations described as stressful are usually, if not always, associated with large changes in secretion of particular hormones.

2 An increased risk of a number of pathological conditions (for example, gastric ulceration and heart disease) is associated with stress. Stress might be described as the state that predisposes to a class of such diseases.

3 Behavioural abnormalities are often associated with stress. In animals, repetitive, apparently pointless behaviours (often self-destructive) are engaged in, sometimes for hours on end. For instance, caged calves often suck urine from the penises of penmates. Humans sometimes engage in repetitive and seemingly pointless behaviours (e.g. self-mutilation by prisoners in solitary confinement, hair-pulling by people with some neurotic disorders) under conditions that are described as stressful.

When you try to abstract what various situations described as stressful have in common, it is not immediately obvious. However, on reflection, the common feature to emerge is that in each case the subject is exposed to a situation for which there is some hindrance to performing appropriate action to alleviate the situation. The frustrated driver on the M25 can do little or nothing to change the situation. The frustrated sleeper cannot terminate the noises. The student with two essays to write by tomorrow can in principle take appropriate action, but stress enters the picture when he or she assesses that the action will be inadequate or other demands compete for time. The restrained animal engaging in rituals has long given up attempts to extricate itself from the situation. The expression 'inability to alter the situation' seems appropriate in each case.

Expressed in these terms, stress would appear to have the features of a negative reinforcement situation (Book 1, Section 6.3.4) in which the action taken fails to extricate the animal from the aversive situation. In stress research an animal having a response available that can terminate an aversive condition is said to have a **coping response** available.

The organizing theme that will be used throughout this chapter is that stress is the state of a living system when it is challenged over a protracted period of time and is unable to take action that is appropriate to the challenge. John Archer at Lancashire Polytechnic employs a similar definition of stress as 'the prolonged inability to remove a source of potential danger, leading to activation of systems for coping with danger beyond their range of maximal efficiency'.

The two key features of stress are: (1) that an individual is placed in a situation that it would not be in if action were able to be taken, but such action to extricate the individual from the situation is not possible, and (2) the danger or threat is protracted. To illustrate the last point, consider an antelope grazing. It detects the presence of a threat, e.g. a leopard coming uncomfortably close. The antelope's body reacts physiologically (e.g. its heart rate rises) and the animal takes appropriate action by running away. It grazes elsewhere. Behavioural scientists would not want to call the antelope stressed. If, however, the same antelope were exposed to a situation over days where it could only drink from a water-hole permanently occupied by carnivores, one probably would want to describe it as stressed, because it cannot take appropriate action.

5.1.2 A model of stress

This chapter will develop a model that is one possible way of looking at stress and provides an organizing framework for the experimental data. This is not to say that the model that emerges is the only one that can be applied to stress. It must be emphasized that there is not a consensus in the literature on how stress should be conceptualized. The model presented here involves just one possible way of giving a definition.

To develop the model, it is useful to look at both the unstressed animal and where the system changes in the conditions described as stressful. In earlier parts of the course (e.g. Chapter 7 of Book 1, and Chapter 4 of Book 5), aggression and fear have been described. In response to a threat, an animal will normally take some kind of action. This might consist of attack or retreat from the situation. For example, in response to a loud bang, birds will tend to take to the air. Rats will tend either to flee to their burrows or, if such escape is not possible, will freeze. Primates take evasive action in response to the presence of higher ranking animals.

In some of the examples just given, the reactions appear rather simple and stereotyped (e.g. a rat freezing upon detecting a predator). In other cases, for example in primate colonies, the reactions can involve some rather sophisticated cognitive processes (Book 1, Chapter 8), in which the action of a threatened animal (e.g. making appeasement gestures or practicing deceit) depends upon the status of the other animals and their anticipated future moves. Whatever the behaviour, the question can always be asked as to how successful a particular behaviour, or strategy, has been. A Siamese fighting fish (*Betta splendens*) that sees off an intruder could be described as having succeeded in eliminating a threat. A rat that instantly flees to the safety of its burrow is successful in staying alive and not getting eaten. It has removed itself from the threat. A member of a primate troop that moves away or displays some kind of an appeasement gesture in the face of a higher ranking animal has succeeded if attack is avoided. If a strategy fails to remove the threat, then an alternative strategy is sometimes implemented. For example, in Siamese fighting fish, if aggressive behaviour fails, withdrawal behaviour tends to be performed (Book 1, Section 7.3).

In each of the situations just described, there is some kind of intrusion which the animal perceives as a threat to its own physical integrity or to its resources. It takes action in response to this threat. If successful, the action changes the situation so as to eliminate the threat. In other words, these situations represent examples of the action of negative feedback systems (see Figure 5.1).

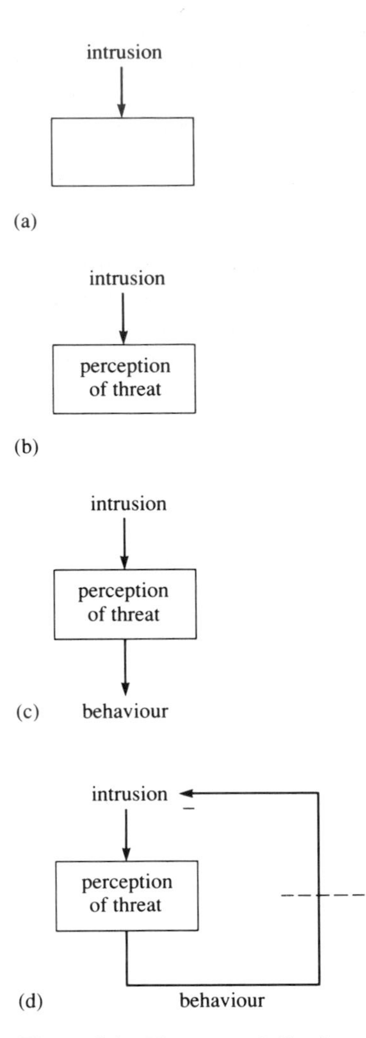

Figure 5.1 The events following an intrusion that is perceived as a threat. (a) intrusion occurs, (b) the intrusion is perceived as a threat, (c) behaviour appropriate to the threat occurs, and (d) if successful, the behaviour corrects the intrusion and eliminates the threat. See text for details.

In response to the perception of a threat, not only is there a behavioural reaction, but there are also profound changes in the physiology of the body. For example, as is common knowledge, human heart rate increases. Changes can occur in the activity level of the gut at times of fear, as is vividly expressed in at least one well-known English idiom. As was discussed in Book 2, Section 8.5, these changes are part of the reaction by the autonomic nervous system (ANS). In addition, another neurohormonal system, termed the hypothalamic–pituitary–adrenocortical system, which is involved in the increased metabolism of sugars, is activated. This expression is quite a mouthful but try not to be intimidated by it; the meaning will shortly be explained. Both the autonomic and hypothalamic–pituitary–adrenocortical changes facilitate the animal's attempts to cope with the threat. For example, by increasing the heart rate, more blood is pumped to the tissues, meaning a greater availability of oxygen and sugars for the muscles. This helps the animal in its fight-or-flight behaviour. Thus, in Figure 5.1 the behavioural action that serves to counter the intrusion depends upon intrinsic (autonomic and hypothalamic–pituitary–adrenocortical) changes.

Using the approach to stress, of looking at the behavioural system and comparing its effective (when correcting the situation) and ineffective (when not correcting it) functioning, a useful working definition of stress is as follows.

1. An animal is exposed to a threat that would normally call for action to correct it.
2. The animal either is prevented from taking action or takes action that is ineffective.
3. As a result, the threat remains.

The term 'stress' is used to characterize the occurrence of situations 1–3 over a protracted period. The expression **stressor** is used to describe a stimulus (e.g. a dominant and aggressive conspecific, or a physical constraint) that evokes stress.

This definition of stress requires some qualification, since humans can often be described as stressed under conditions that are not exactly threatening. However, it is a useful starting point to bring some order to the area of stress. It fits well with the other criteria of stress listed earlier. In the state described as stress, action is taken but this action fails to have the effect of removing the intrusion. In terms of Figure 5.1d, the link is broken at the point marked by the broken line. This is a situation described by the expression **open loop**. If a threat results in action that removes the threat, the system is said to exhibit a **closed loop**. The term 'stress' covers both the psychological and physiological states that are associated with this failure to close the loop. Failure to remove the threat is associated with abnormal levels of secretion of hormones, an increased tendency to disease of various sorts (pathology), and abnormal behaviour. Long-term stress is often described in terms of the harm that is caused to the body by elevated hormone secretions; however, there might also be some advantages that such secretions serve, as will be discussed later.

The action taken in response to an intrusion can take various forms. Attack and escape are two obvious possibilities. In some species (e.g. rats, guinea-pigs), immobility (sometimes termed freezing) is another possibility. Like attack and escape, immobility can either succeed or fail. As a strategy to avoid predators freezing makes good sense, as by remaining immobile the animal is less conspicuous. Confronted with conspecifics, freezing can also be of adaptive value. Mice and rats are less likely to be the targets of aggression if they adopt the 'appeasement gesture' of freezing. During the period of immobility, hormonal changes can still prepare the animal for fight-or-flight, should this turn out to be necessary. In the following sections, it will be argued that stress arises when either active (e.g. fighting and fleeing) or passive (e.g. immobility) strategies fail.

In order to understand stress, one possible starting point is to look at the normal physiological response underlying the actions of fight and flight, which forms the topic of the next section. In these terms, stress is associated with placing abnormal demands upon this physiology.

5.2 The physiology of fight-or-flight

In response to the appearance of a threat, there are numerous changes that occur in the physiology of the body. These changes will both *affect* and *be affected by* the subsequent behaviour. Levels of a variety of hormones and neurotransmitters will change, as will the activity of various neural pathways. However, researchers into stress have traditionally found it convenient to focus attention on a particular set of changes that play the most obvious role, within the ANS and in the hypothalamic–pituitary–adrenocortical system. Each of these will be considered in turn.

5.2.1 The autonomic nervous system

This system was described in Book 2, Chapter 8. The autonomic nervous system (ANS) is the part of the nervous system that controls the activity of the organs of the body, e.g. heart rate and the contractions of the intestine depend upon its activity. The term 'autonomic' reflects the fact that the system is to some extent self-governing. The ANS keeps the organs functioning even in the absence of conscious control; movement of material through the gut proceeds without one having to think about it, for example.

However, there are important connections between the ANS and the brain regions outside the system (e.g. cortex). For example, the perception of a threat through the sensory system and processing within the cortex leads to activation of the ANS. This input gives rise to an integrated activity throughout the whole of the ANS to meet challenges to the whole animal. It is this integrated mode of activity of the whole ANS that will concern us here.

There are two branches of the ANS, the sympathetic nervous system (SNS) and the parasympathetic nervous system (PNS). As a general rule, increased activity in one of these branches has the opposite effect to increased activity in the other. For example, whereas increased sympathetic activity leads to increased heart rate, increased parasympathetic activity leads to decreased heart rate. Furthermore, an increase in activity of one branch of the ANS is generally accompanied by a decrease in activity in the other branch. In response to a challenge for which fight or flight is the appropriate response, the sympathetic branch will be activated.

Suppose an animal is exposed to a predator and immediately either attempts to launch an attack or takes the evasive action of running away. Typically, on the first encounter with the predator, the activity of the SNS will increase instantly, heart rate will increase and, as a result, the supply of blood to the tissues increases. Blood vessels in the gut wall will constrict and, as a result, blood will be diverted from there, thus slowing down the process of digestion.

☐ What is the functional significance of this shift of blood flow?

■ When time is at a premium, digestion can wait. The shift of blood from the gut makes more blood available to the muscles where its immediate availability (as a source of oxygen and nutrients) is imperative for survival if exertion is called for.

What is the mechanism of action of the SNS? There are two aspects to this, direct and indirect. The neurons of the SNS innervate smooth muscle and heart muscle. In response to activity in the SNS, the neurotransmitter noradrenalin is released at nerve-muscle junctions. On binding to receptors at the muscle, this changes the state of the muscles. For example, heart muscle is made to contract faster. This is the *direct* action.

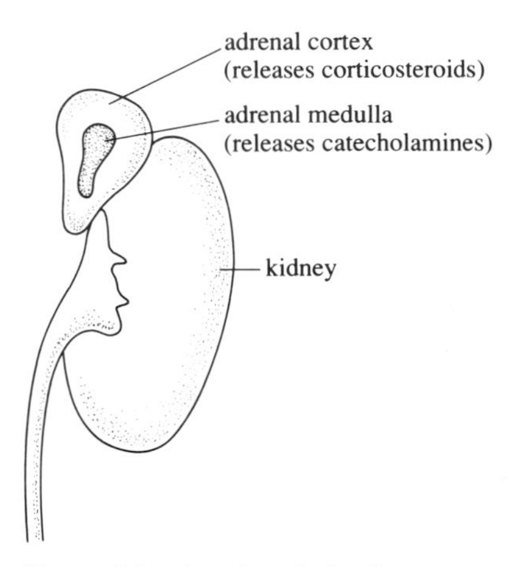

Figure 5.2 An adrenal gland.

The second, *indirect*, aspect arises as a result of neurons in the SNS innervating the adrenal glands (Figure 5.2). There are two adrenal glands, one above each of the two kidneys. Activity in the SNS causes the release of two hormones from the adrenal glands, adrenalin and noradrenalin (also called epinephrine and norepinephrine, respectively). More specifically, their release is from the centre of the glands, termed the adrenal medulla. These two hormones are then carried by the bloodstream around the body. Adrenalin and noradrenalin are members of a class of substance termed catecholamines (Book 2, Section 4.5.2) and in the context of stress are often referred to by this generic expression. Figure 5.3 shows a summary of the processes described so far.

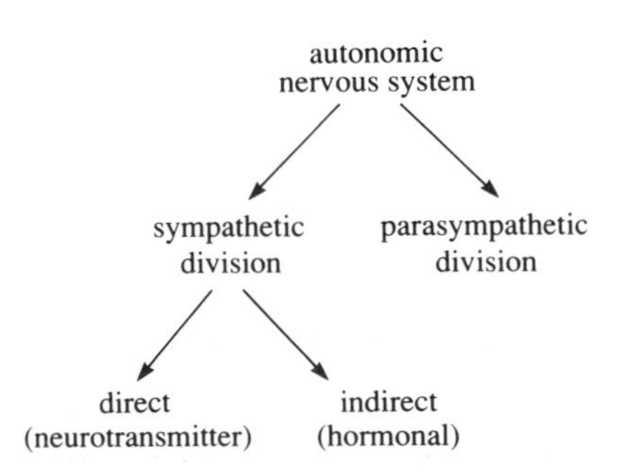

Figure 5.3 Summary of the processes in the ANS.

☐ Why is noradrenalin sometimes termed a hormone and sometimes a neurotransmitter?

■ There is just the one substance. The terminology refers to two different roles of the same substance. At those sites where noradrenalin is released from a nerve terminal, diffuses across a synaptic cleft and influences an adjacent cell, it is termed a neurotransmitter. Where it is released by the adrenal medulla and then transported around the body in the blood, to act at a variety of different sites, it is termed a hormone.

At a particular site, the amount of contraction shown by a muscle depends upon the amount of noradrenalin and adrenalin present at receptors on the muscle. This will be made up of noradrenalin released locally by neuron terminals, and that derived more diffusely via the bloodstream.

□ Is the function of the adrenal medulla to play a role in effecting local responses by the SNS or responses throughout the body ('global responses')?

■ Since adrenal hormones are released into the bloodstream and carried around the body, activation of their release will have global effects, i.e. they will act upon all those organs that have appropriate receptors. There is no way that released hormones can have specific local effects on only some of the organs with appropriate receptors.

If a behavioural strategy is successful in that, for example, either the predator flees or the potential prey manages to reach a safe haven, the input to the SNS is lowered. Activity in the neurons of the SNS falls and, as a result, the rate at which noradrenalin is released from nerve terminals in the SNS drops to normal (or, as it is often termed, 'basal') levels. Similarly, release of hormones from the adrenal medulla will fall to basal levels. Note that adrenalin and noradrenalin are being broken down in the body all the time. If the level of a hormone is observed over a period of time to be constant, this implies that it is being broken down as fast as it is being released into the bloodstream.

□ What is implied if the concentration of a hormone in the blood is (a) rising, or (b) falling.

■ (a) The rate of release into the bloodstream is greater than the rate of breakdown.
(b) The rate of breakdown is greater than the rate of release.

Without the elevated sympathetic input to the adrenal medulla, the elevated levels of adrenalin and noradrenalin concentration in the blood will not be maintained. The natural outcome of removing SNS activation is that their levels will fall back to normal.

Suppose though that the animal makes efforts in response to an intrusion and these are not successful. Of course, one possible outcome is that it gets eaten, but that is not our prime concern here! Consider a less drastic but more chronic possibility. In a situation where two animals are competing for territory, for example in a laboratory cage, one might become dominant to the other. The other takes evasive action but cannot remove itself far enough away to avoid being attacked by the dominant. The SNS of the animal taking evasive action would be activated on first encounter but, rather than returning to the base-line, would typically remain chronically elevated, with a correspondingly elevated heart rate. This is a state that would be characterized as stressful according to the model developed in this chapter. It involves stretching an adaptive control system to beyond its limits of optimal functioning, and it increases the risk of pathology.

5.2.2 The hypothalamic–pituitary–adrenocortical system

Also of central importance in studying stress is another class of hormone, known as corticosteroids, so-named because they are released from the adrenal cortex, the outer layer of the adrenal glands (Figure 5.2). Corticosteroids are also referred to as glucocorticoids since they have an effect on glucose metabolism. There are a number of different hormones in this class having similar properties; one that you will meet later is cortisol. These hormones promote the mobilization of energy reserves, such as making glucose available in the bloodstream, and play a crucial role at all times, whether the animal is under stress or not, in maintaining the energy balance and metabolism of the body's tissues.

On application of a stressor, there is an increased release of corticosteroids, which mobilize energy-providing substances from reserves and thereby increase the availability of glucose to the tissues. In combination with the action of catecholamines, discussed in the last section, corticosteroids therefore play a fundamental role in providing for a level of metabolism that will allow the action of fighting or fleeing. On close scrutiny, the sequence of events that leads to the release of corticosteroids is complex, but do not be discouraged since a somewhat simplified version of the essentials of this sequence will now be described, starting in the brain.

Figure 5.4a shows certain neurons in the hypothalamus that synthesize a hormone, known as corticotropin-releasing hormone (CRH). They release this hormone into blood vessels that transport the CRH directly to the pituitary gland (see also Figure 5.4b for a summary of the sequence). The CRH-containing neurons in the hypothalamus receive excitatory inputs from other regions of the brain, indicated by plus sign in Figure 5.4b. Detection of a threat causes activation of neurons that impinge upon the CRH-containing neurons, causing them to increase their release of CRH.

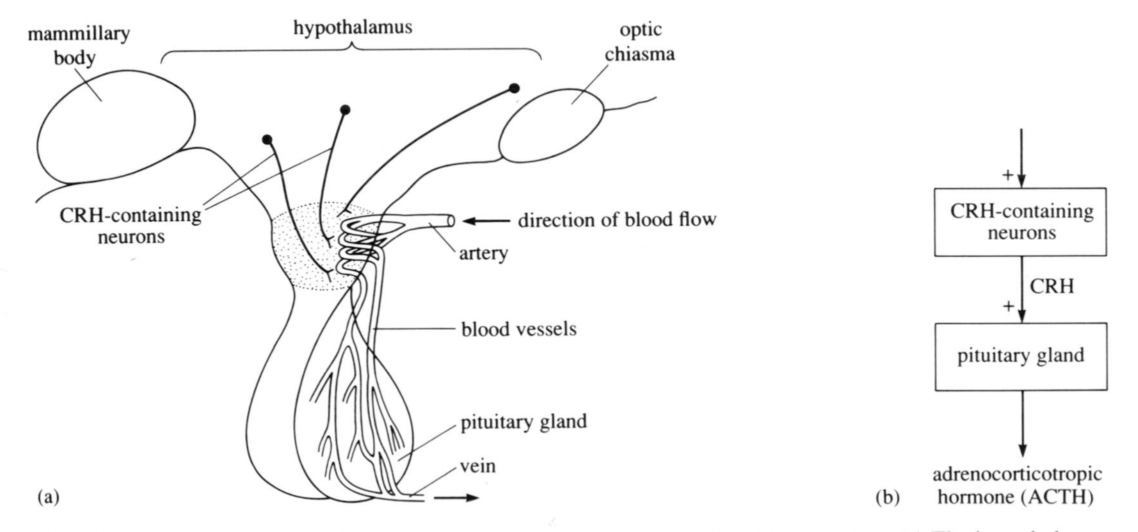

Figure 5.4 The initial components of the hypothalamic–pituitary–adrenocortical (HPA) system. (a) The hypothalamus and pituitary. Note the CRH-containing neurons which, when excited, pour their contents into blood vessels which transport the CRH to the pituitary. The pituitary then stimulates the release of ACTH into the circulation. The dotted area shows the area of the hypothalamus where CRH secretion takes place. (b) The sequence of events in the hypothalamus and pituitary.

At the pituitary, the effect of CRH is to stimulate the release of another hormone, adrenocorticotropic hormone (ACTH) into the bloodstream. This hormone is then distributed widely throughout the body. Its site of action is the adrenal cortex, where it promotes the release of corticosteroids. The physiology and its sequence of actions just described constitute the hypothalamic–pituitary–adrenocortical (HPA) system (sometimes called the HPA axis instead of the HPA system). The sequence of events is summarized in Figure 5.5.

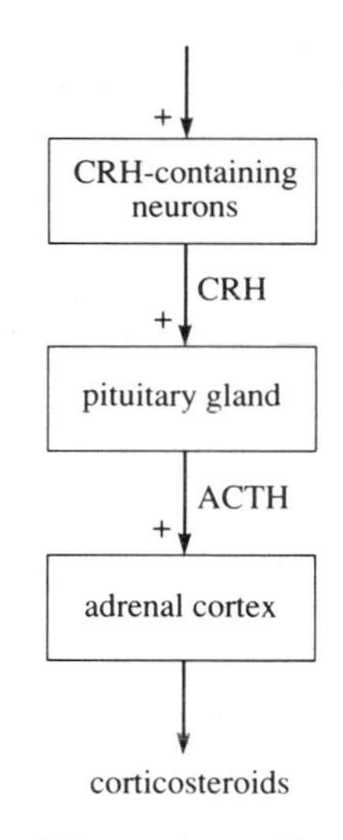

Figure 5.5 Excitatory factors leading to the release of corticosteroids.

Figure 5.5 shows the excitatory factors in the HPA system that lead to a rise in blood corticosteroid levels. Two principal factors limit this rise. First, as well as excitatory factors affecting the HPA system, there are also inhibitory factors, as shown with minus signs in Figure 5.6. These inhibitory influences act both upon the CRH neurons and the pituitary. Inhibition at either site causes a reduction in the release of ACTH and thus a reduction in the release of corticosteroids. Corticosteroid receptors are found at various sites in the brain, e.g. the hypothalamus and hippocampus, as well as in the pituitary. When these sites are occupied by corticosteroids, inhibition is exerted.

An increased release of CRH causes the increased release of ACTH, which *increases* the release of corticosteroids. These then occupy sites that cause inhibition to be exerted, which *reduces* the rate of release of corticosteroids.

□ What type of system is this?

■ A negative feedback system (Book 1, Section 7.2.2).

Secondly, as with other hormones, there is a process of breakdown. If a constant level of hormone in the blood is observed over a period of time, you will recall that this implies that the hormone is being broken down as fast as it is being released into the bloodstream.

When a stressor is applied, the system will be activated. First, excitatory neurons that synapse upon CRH-containing neurons are activated. CRH triggers the release of ACTH. In response, corticosteroid levels in the blood will rise. As corticosteroid level rises, so the negative feedback effect will increase. This tends to restrain the increase in the release rate of corticosteroids. Figure 5.7 (*overleaf*) shows the sequence of events.

□ In Figure 5.7, after time t_2, the level of corticosteroids is constant. What does this imply about their release and breakdown?

■ The breakdown rate of corticosteroids is equal and opposite to their release rate from the adrenal cortex.

There is also a circadian rhythm (Chapter 3) in corticosteroid levels. Even under constant conditions with no application of a stressor, a rise and fall in corticosteroid level is observed over a 24-hour period. This arises because of a rhythm in neural input to the CRH neurons in the hypothalamus. Note that the effect of a stressor acts against this background level of activity within the HPA system, repesented in Figure 5.7c by the basal level of secretion of corticosteroids. It is not that the system lies inactive and a stressor triggers it into activity. Rather, a stressor increases the activity of the system to above its basal level.

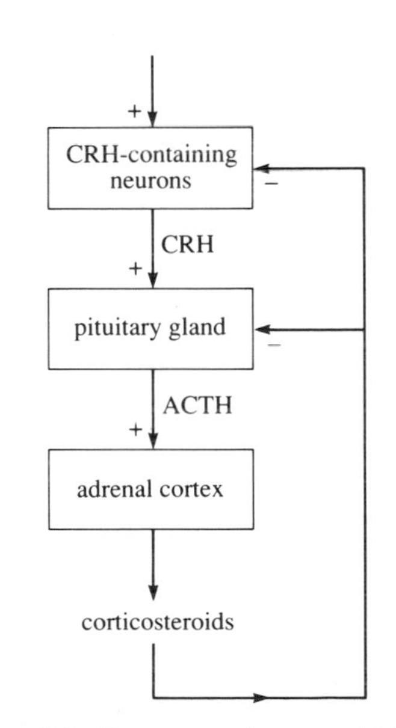

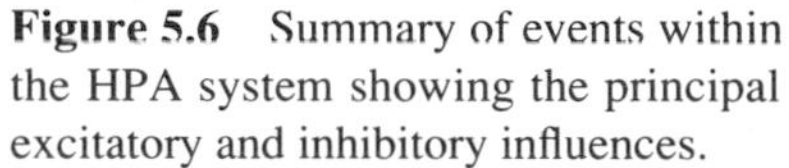

Figure 5.6 Summary of events within the HPA system showing the principal excitatory and inhibitory influences.

Thus, in response to a stressor, both the sympathetic and HPA systems are activated. Amongst other things, their activation enables the metabolic changes to occur that facilitate fleeing or fighting. If the situation is a closed loop as shown in Figure 5.1, this action succeeds in placing a distance between the animal and the stressor. The stressor is removed, and so the excitatory inputs to these two neurohormonal systems are removed. The systems can then return to their basal levels.

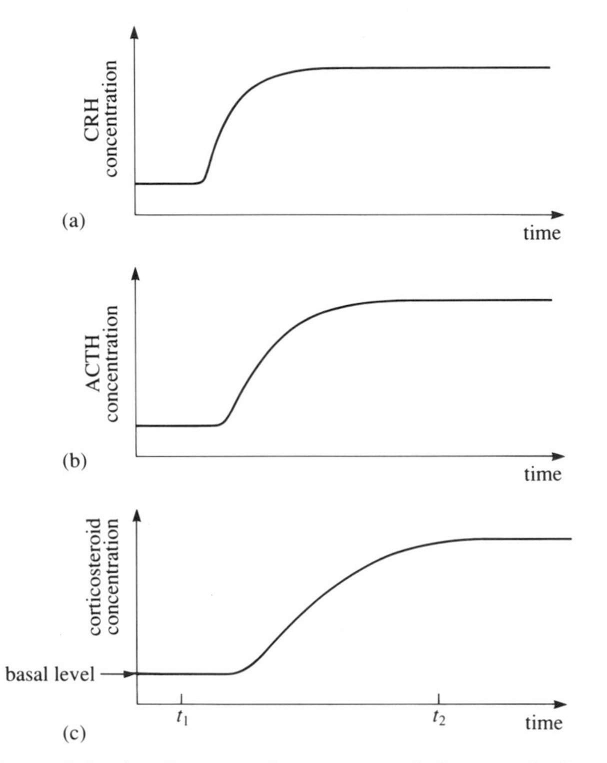

Figure 5.7 Events following the onset of a stressor applied constantly from time t_1: (a) the release rate of CRH into vessels linking the hypothalamus and the pituitary increases, (b) there is a rise in blood ACTH concentration, and finally (c) there is a rise in blood corticosteroid levels.

□ For two reasons, this outcome is not one that would be characterized as stressful. What are they?

■ First, the metabolic changes are appropriate to the action taken. Second, the action is successful. On successful completion of the action, both neurohormonal systems return to their basal levels.

Stress occurs when action is taken but is in some way ineffective in restoring normal conditions. This is associated with metabolic changes that appear inappropriate to the behaviour that the animal performs. For example, there could be a situation in which there is activation of the HPA system, which would be appropriate for the increased metabolism of activity, but all the animal can do is show passivity. This argument could be applied to a number of different research subjects, but is perhaps best illustrated by some classic studies on the tree shrew (*Tupaia belangeri*), the topic of the next section.

5.3 Stress in tree shrews

In their natural habitat, tree shrews form pairs and fight to defend their territory against intruders. They are found in the forests of Thailand and appear something like a slightly smaller version of the familiar grey squirrel. Dietrich von Holst (1986) has studied the behaviour of captive tree shrews at the University of Bayreuth in Germany.

Before the start of the experiment, animals were housed individually, with some animals being implanted with small radiotransmitters to broadcast their heart rates. At the start of the experiment, two male tree shrews met for the first time on being placed together in an unfamiliar cage equipped with two separate sleeping boxes. Their subsequent behaviour was monitored closely and, at intervals, their blood was sampled to investigate hormone concentrations.

After the two animals were first introduced to the cage, they explored and scent-marked it. After a short period of time, fighting began. Within 1–3 days, this fighting lead to the emergence of a clear dominance relationship. Subsequently, the winner, termed the dominant, would, for the most part, ignore the loser, with attacks on it being rare or absent. However, the behaviour of the loser was profoundly affected. On the basis of their distinct reactions to the dominant, von Holst identified two sub-groups of losers, which he termed *subdominant* and *submissive*.

The behaviour of the subdominants was characterized by *active* attempts to avoid the dominant. This involved continual monitoring of the moves of the dominant and avoidance of confrontation by giving way. On the rare occasions when they were attacked, the subdominants would attempt to defend themselves. By contrast, the behaviour of the submissives could be characterized as *passive*, or as giving up. They crouched in a corner of the cage or in a sleeping box, and would leave their sanctuary only rarely to eat and drink, which was done rapidly. On occasions when attack from a dominant was directed at them, they would make no attempt to either fight or flee. Submissives ceased grooming and their coats became distinctly dirty. Observers used the terms 'apathetic' and 'depressed' to describe their appearance. As Figure 5.8 (*overleaf*) shows, dominants, subdominants and submissives can be clearly distinguished by their weight following the encounter.

Of the hormone concentrations monitored (testosterone, corticosteroids, adrenalin and noradrenalin, i.e. those considered most relevant to stress), no difference was found between dominants, subdominants and submissives at the start of the experiment.

□ What can we conclude from this observation?

■ That the different behavioural reactions shown could not be attributed to *initial* differences in these hormone levels.

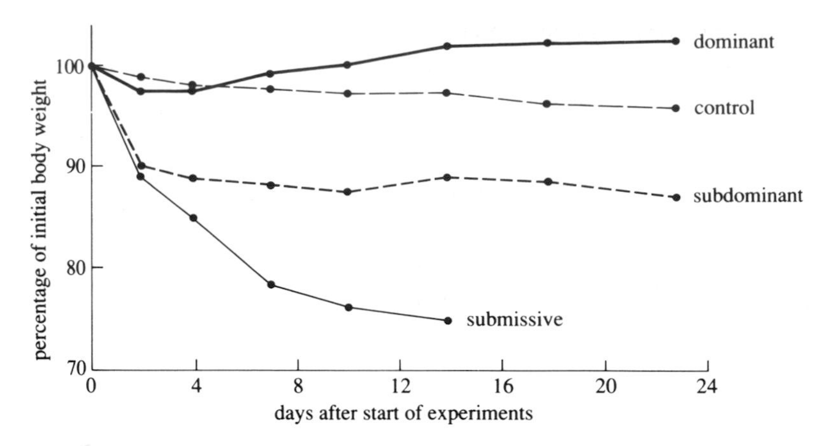

Figure 5.8 Changes in weight of tree shrews following the start of the encounter. Controls were animals living alone in their home cages.

Two types of corticosteroid were monitored, cortisol and corticosterone. For all animals, levels of corticosteroids in the blood were elevated during days 1–3 of confrontation. Figure 5.9 shows the levels of these two hormones after 10 days of confrontation, relative to controls.

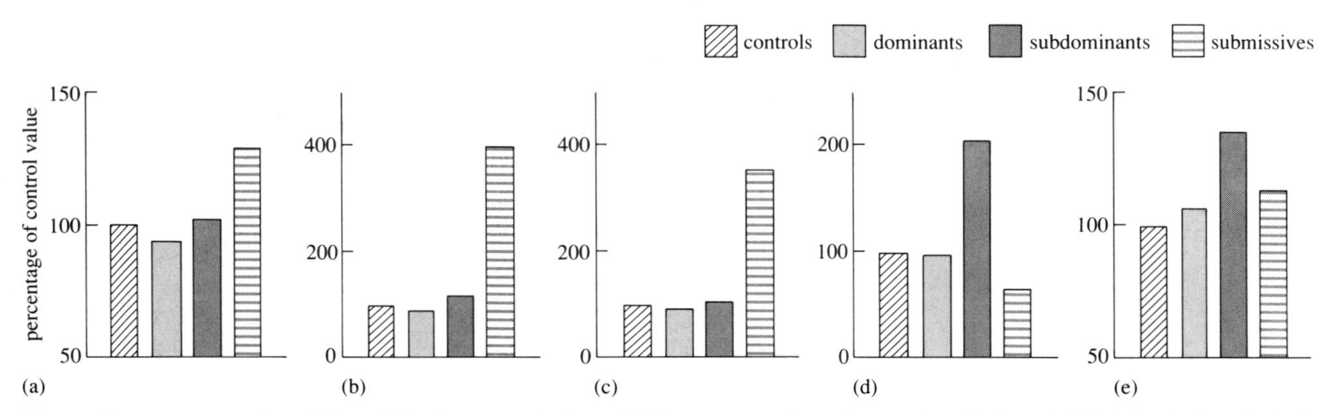

Figure 5.9 Measures of variables within the sympathetic and HPA systems of tree shrews after 10 days. (a) Weight of adrenal glands, (b) level of cortisol in the blood, (c) level of corticosterone in the blood, (d) adrenal content of the catecholamine synthesizing enzyme, tyrosine hydroxylase, and (e) adrenal noradrenalin content.

□ What do the results show as far as corticosteroid levels are concerned?

■ They show that, for both dominant and subdominant animals, after 10 days, corticosteroid levels (i.e. both cortisol and corticosterone) have returned to normal. For submissive animals, they are still grossly elevated.

□ By the criterion of appropriateness of hormonal levels, would the submissives be said to be stressed?

■ Yes. The elevated corticosteroid levels are appropriate for the increased metabolism of fight-or-flight. They are inappropriate for the inactivity of the submissive animals.

Notice also the elevated weight of the adrenal glands in the submissives relative to the other animals. Figure 5.10 shows the heart rate of the dominant and subdominant animals prior to and following the initial encounter. In both cases a clear circadian rhythm (Chapter 3) is evident.

☐ What is the main difference in the response of the dominants and subdominants?

■ Following the initial encounter, the heart rate of the dominant is briefly elevated but quickly returns to the base-line. The heart rate of the subdominant is elevated and remains so.

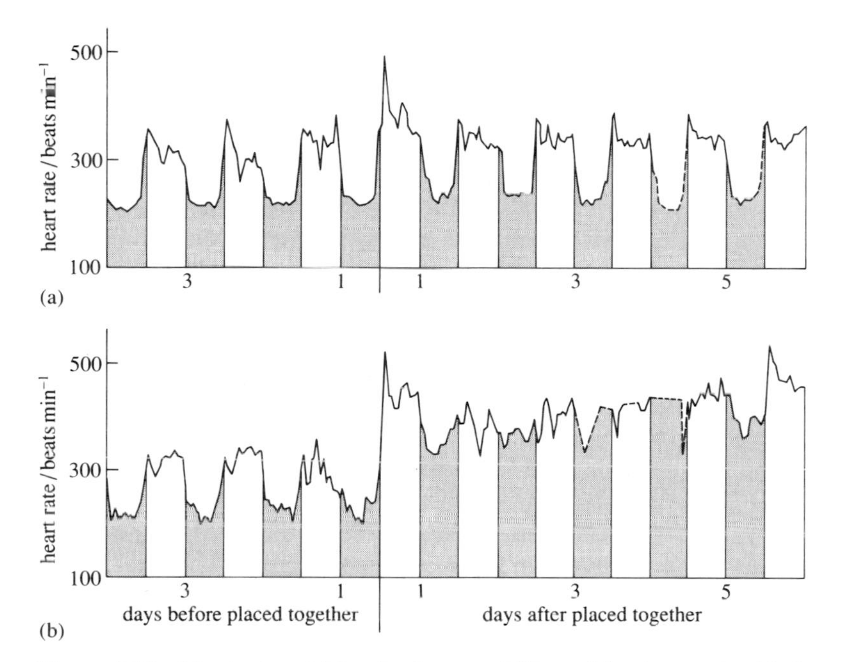

Figure 5.10 Heart rates of (a) a dominant and (b) a subdominant animal before and after being placed together. The shaded areas indicate darkness, and the unshaded areas indicate light.

As can be seen, the subdominant's heart rate is elevated even at night, when both animals are sleeping in separate boxes.

☐ What does the heart rate response suggest regarding the ANS?

■ That the SNS is activated following the encounter, in a transient way for the dominant and in a sustained way for the subdominant.

Unfortunately, von Holst was not able to obtain reliable heart rate recordings for submissive animals since their transmitters tended to be rejected, but he reported his impression '...that in contrast to the active subdominants, submissives have a reduced heart rate and lowered blood pressure (evident during sampling)...'. At autopsy, the heart weight of submissives was found to be 20% lower than that of other groups.

If there is a difference in SNS activity between different types of animal, this might be expected to be mediated, at least in part, by a difference in the release of adrenalin and noradrenalin from the adrenal medulla. This would imply differences in the rate of synthesis of the hormones. Any such difference might be revealed in different levels of the enzymes that play a role in the synthesis of these hormones within the adrenal glands. One such enzyme is tyrosine hydroxylase (TH). The adrenal content of this was no different from control values for the dominants but was elevated by more than 100% for subdominants. It was reduced by about 30% in submissives (Figure 5.9d). A significant elevation of adrenal noradrenalin (Figure 5.9e) was seen in subdominants but not in other animals.

Von Holst concludes that, following their initial encounter, one of two possible strategies will in time emerge from the loser, each accompanied by a particular neurohormonal change. For animals designated as subdominant, there is an *active strategy* accompanied by increased and sustained activity of the SNS. No lasting change in HPA activity was seen beyond the first few days of the experiment. For animals designated as submissive, there is a *passive strategy* accompanied by increased activity of the HPA system that is sustained for the duration of the experiment. There is a tendency to reduced sympathetic activity in submissives.

On the basis of his tree shrew experiments, von Holst suggests that there are two types of stress, each with an associated predisposition to pathology. The heightened heart rate of the subdominant predisposes it to pathology of the circulatory system, e.g. cardiac failure. For the submissives, von Holst predicted an impaired immune system (discussed later) and a reduction in wound healing ability as a result of chronically elevated corticosteroid levels.

Studies on a range of other species, including mice, rats, monkeys and humans, point strongly to a similar classification of two types of stress. It seems that a given animal has two possible strategies available to it: passive or active. According to the context, one will emerge and one will be suppressed. However, if one strategy fails, there is a chance that the other can still be 'switched on'. In particular, if an active strategy fails, the passive will emerge. Thus, a given animal might be thought of as having a bias towards one or other strategy, but which strategy emerges depends upon context. Research has shown clearly that selective breeding (Book 1, Section 3.5.2) can create strains of rats or mice with a bias towards one or other strategy. Some examples of the pathology associated with prolonged experience of one or other system being dominant form the topic of Section 5.5. The next section looks at another possible index of stress: the performance of abnormal behaviour.

5.4 Repetitive behaviour

Earlier it was noted that one of the pieces of evidence commonly used for the existence of stress is the performance of abnormal behaviour, usually highly repetitive, stereotyped behaviour over prolonged periods of time. Such so-called *stereotypies* commonly appear at a stage where other attempts by the animal to extricate itself from an aversive situation have failed. For instance, the first reaction of a sow to tethering is one of attempted escape. This is followed by signs of aggression and finally by stereotypies, such as repetitive chewing of a bar.

The motor acts of licking, sucking and chewing have a variety of hormonal effects. For example, the plasma levels of somatostatin, a hormone that appears to be released under conditions of stress, is lowered by sucking. Some of these hormonal changes might have implications for the animal's physiology. Piet Wiepkema working in the 1980s at the Netherlands Agricultural University in Wageningen found that veal calves that showed the behaviour of repetitive tongue playing had no gastric ulceration. Those housed under identical conditions but not developing this habit showed clear evidence of gastric ulceration.

□ From this evidence, why would one need to be careful in concluding that tongue playing had beneficial effects on the animal's physiology?

■ A negative correlation between tongue playing and gastric ulceration was found. A correlation on its own does not permit one to infer that the one thing caused the other (see Book 1, Section 2.4.2).

Although this experiment does not enable one to conclude that tongue playing has beneficial effects, it points to such a possibility. It also serves as a warning as to how important individual differences between animals can be.

There is some evidence that performing stereotypies serves to lower or restrain activity levels in the HPA system. For instance, caged voles show a species-typical jumping stereotypy. If this response is thwarted by lowering the ceiling of the cage, a rise in corticosteroid levels is observed. A number of theorists have speculated that stereotypies serve to trigger the release of endorphins and some experimenters have found a sharp reduction in the performance of stereotypies following injection of naloxone, an opiate-antagonist. However, at the time of writing, there is conflicting evidence and no consensus on their role, though this topic is subject to a massive research effort.

5.5 Stress and disease: application to human pathology

5.5.1 Circulatory system pathology

Under the influence of adrenalin and noradrenalin, amongst other things, substances known as free fatty acids are released into the bloodstream. These form substrates from which energy can be derived. Thus, sympathetic activation leads to an elevation of such free fatty acids. Both physical exertion (e.g. jogging, aerobics) and stress are associated with sympathetic activation. However, in the case of exercise, the free fatty acids are utilized as an energy source and are thus cleared from the circulation. If they are not cleared, as in the case of, for example, a driver stressed by being stuck in the London traffic, they can, after some chemical conversions, form deposits—plaques—of fatty material on the walls of the arteries (Figure 5.11, *overleaf*), a condition termed *atherosclerosis*. High blood pressure, another consequence of elevated sympathetic activity, increases the rate of formation of such plaques. Elevated catecholamine levels in the blood also appear to play a role in plaque formation. In this way, several factors associated with elevated sympathetic activity can act together in creating a pathological state. Diet acts together with such factors in promoting this condition. In studying heart disease in humans, it is very hard to disentangle the effects of stress from those of smoking, obesity and poverty, as well as the role of the genotype.

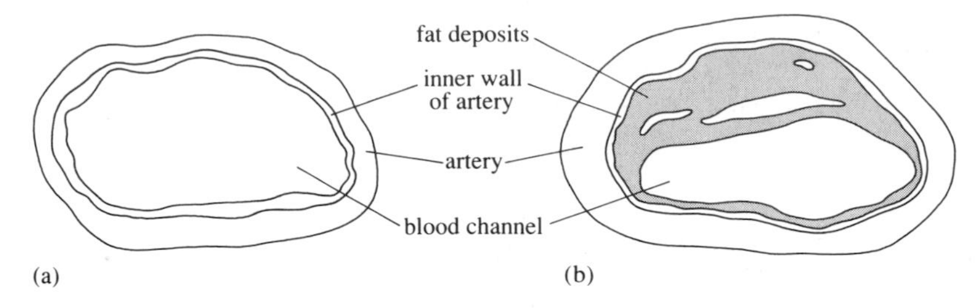

Figure 5.11 Drawing of a cross-section through (a) a healthy and (b) an atherosclerotic human artery. Note the fatty deposits (plaques) attached to the inner wall, reducing the width of the channel.

In human studies, it has been suggested that personality may be one factor amongst many others associated with circulatory disorders: a so-called 'coronary-prone personality' is frequently described. Though the area is fraught with controversy, this personality type is usually associated with impatience, hostility and excessive competitiveness. However, some researchers more closely associate such traits as tension, neuroticism and anger suppression with coronary heart disease, whereas others find no such association at all.

Studies on non-human primates have enabled some conclusions to be drawn in this area. In studies on baboons, the highest levels of cholesterol in the blood were found in animals taking active coping strategies that were not successful. They might be described as 'upwardly struggling' members of the population. In a study of male crab-eating macaque monkeys (*Macaca fascicularis*), researchers found most atherosclerosis to be associated with dominant males living in a socially unstable situation where their dominance was under threat. A combination of social instability and a high cholesterol diet was particularly effective in promoting atherosclerosis.

Though doubtless the level of activity of the HPA system also plays a role in disorders of the circulatory system, traditionally the focus of attention has been on elevated sympathetic activity. Such disorders appear usually to be the result of a chronic failure of active strategies to change the situation.

5.5.2 Predictability and controllability of the stressor and pathology

One of the pathological conditions most closely associated with stress in the mind of the general public is gastric ulcers. Experimentation has cast important light on some of the factors that determine ulceration. The experiments described here are particularly likely to raise in your minds important issues concerning the ethics of animal research (discussed on the audiotape *The Use of Animals in Research*). They raise a very real and painful dilemma for many people working in this area, since their scientific importance is beyond doubt and yet they would appear to involve inflicting clear suffering upon the experimental subjects.

In one study, it was shown that the physical property of the stressor to which a subject is exposed is only one of the factors that determines its impact. Another crucial factor is the **controllability** of the situation. Rats were tested in groups of three, two of which were subjected to electric shocks applied to the tail. The third was an unshocked control. One rat of each pair, termed the active rat, was able to terminate the shock by turning a wheel at the front of the cage. The other rat, the passive rat, also had a wheel available but turning this wheel had no effect upon the electric shock. The rats were physically separated and could not see each other.

□ Why was a wheel provided for the passive rat when it had no effect upon the outcome?

■ To keep the conditions as similar as possible between animals, except for the *one* variable under observation, in this case the capacity of the rat to influence the shock.

The active and passive rats were 'yoked' in the sense that the same electric shock passed through each of their tails: by terminating the shock to its own tail, the active rat terminated the very same shock to the passive rat's tail. Thus the physical stressor to which they were exposed was identical. The interesting question is whether the effect of the stressor is the same in the two. The active rats had far fewer gastric ulcers and gained weight faster than the passive rats.

The active rats were able to exert some *control* over the situation; their behaviour was of some consequence to what happened. The fact that the response of turning the wheel terminated the shock provided *relevant feedback*. This relevant feedback served in some way as an antidote to the deleterious effects of the stressor. Another common way of expressing this is to say that the active rats were able to acquire a coping response for dealing with the situation.

□ Is it possible to interpret the results and the conclusion in terms of the general definition of stress presented earlier?

■ By its behaviour, the active animal is able to effect change in its environment to terminate the stressor. It has a coping response. The system is a closed loop, i.e. (stressor) → (behaviour) → (eliminate stressor). By contrast, any behaviour by the passive yoked control is without effect upon the stressor: it is summarized simply as (stressor) → (behaviour).

The concept of controllability is very relevant to understanding depression, which forms the topic of the next section.

5.5.3 Control and depression

In recent years, the notion of coping has assumed central importance in studies of stress, both in humans and non-humans. The literature on stress contains numerous references to the desirability of giving people more control over their environments. For example, people living in old people's homes tend to be psychologically healthier if some control is placed in their hands, for example, being involved in the day-to-day running of the establishment, or sharing essential tasks.

A number of theorists have linked depression with lack of control: depression is associated commonly with feelings of hopelessness and helplessness. Depressed people tend to assume an inability on their own part to influence events. A powerful school of thought in psychology identifies stress as a precipitating factor in many cases of depression. Researchers have looked for biochemical changes in the brain that result from exposure to an uncontrollable stressor and which might later predispose the subject to depression.

Suppose animals are first exposed to an uncontrollable stressor and are then tested in a situation where control can be exercised. The prior experience with the uncontrollable stressor means that subsequently they either fail to learn to perform coping strategies in the controllable situation, or are seriously retarded in their learning ability. It is often argued that in the first phase of the experiment the animals learn that their behaviour does not influence events and that this learning carries over to the second phase, a phenomenon termed **learned helplessness**. The converse situation also applies: exposure to a controllable stressor can serve to protect the subject against subsequent deleterious effects of exposure to an uncontrollable stressor.

There is evidence to show that exposure to uncontrollable stressors leads to changes in the levels of neurotransmitters (e.g. noradrenalin, serotonin) at various sites in the brain. Exactly which neurotransmitter change is crucial, or whether a combination is, constitutes a topic of lively debate at the moment. Whatever the exact change that is implicated, a guiding principle for researchers is that some such neurotransmitter change plays a role in depression. Evidence suggests that noradrenalin levels at certain key sites (like the levels of dopamine, discussed in Chapter 2) play a role in the motivational mechanisms of the animal. In the present context, could it be that the depleted brain noradrenalin levels associated with exposure to uncontrollable shock play a causal role in the subsequent failure to show coping responses when control is possible? A certain amount of evidence suggests that this is the case. A depletion of noradrenalin caused by chemical manipulation of the brain has a very similar effect to exposure to uncontrollable shock. Conversely, protecting the brain from noradrenalin depletion (by injecting a drug that prevents its breakdown) during exposure to uncontrollable shock tends to protect against the effects of shock.

An animal characterized by (a) noradrenalin depletion, and (b) not exerting control in the face of adversity might provide a suitable animal model for at least one important aspect of human depression (Book 6). Another important aspect of depression for which stress research might provide an 'animal model' concerns the HPA system and is discussed in the next section.

5.5.4 Depression and the HPA system

In discussing the work of von Holst on tree shrews (Section 5.3), it was noted that losers in a conflict showed one of two types of response: subdominants made active coping attempts, whereas submissives gave up the struggle.

□ What was one of the most striking hormonal changes that characterized the submissives?

■ They showed a greatly elevated level of corticosteroids in the blood.

Evidence from a variety of species, including mice, rats, human and non-human primates, shows that passivity in the face of adversity is associated with elevated corticosteroid levels.

□ From these observations could it be concluded that the elevated activity level of the HPA system caused the passive behaviour?

■ No. The passivity might have caused activation of the HPA system. Another possibility is that there are reciprocal interactions, such that behaviour and mood influence the hormonal state and that the hormonal state also influences behaviour and mood.

In humans with depression, the cortisol secretion rate is often abnormally elevated. In order to conclude that such an abnormal hormonal state can affect behaviour, further evidence is needed. In humans, certain disorders that are characterized by a high cortisol secretion rate, and which have identifiable origins in body tissues outside the nervous system, are associated with depression that is a secondary effect of the hormonal disorder. This suggests that, in subjects whose primary disorder is depression, corticosteroids can play a role in the depressed mood.

□ In terms of an understanding of the sequence of events in the HPA system (Figure 5.6), what might contribute to the elevated levels of cortisol seen in depression?

■ It could be either: (1) an abnormally high level of hypothalamic–pituitary activity, e.g. a high CRH secretion rate, (2) an increased sensitivity of the pituitary to CRH, (3) an increased sensitivity of the adrenal cortex to normal levels of ACTH, or (4) the corticosteroids might break down more slowly.

Evidence suggests that both factors (1) and (2) play a role. One focus of interest is the corticosteroid receptors in the brain.

□ According to the earlier discussion, what role does activation of these receptors play?

■ Their occupation by corticosteroids exerts negative feedback on the HPA system.

A number of studies in various species has shown that, in a condition characterized by 'lack of control', 'helplessness' or 'depression', there is a reduction in the number of corticosteroid receptors in the brain.

□ What would the predicted effect of this be as far as the HPA system is concerned?

■ Normally, negative feedback is mediated via occupation of the receptors by corticosteroids. If the receptor numbers are reduced, inhibition will be reduced. Elevated corticosteroid secretion is therefore predicted.

Indeed, the reduction in the number of corticosteroid receptors appears to be a crucial link in explaining the drastic rise in corticosteroid levels. There is evidence that high levels of corticosteroids have a toxic effect upon their own receptors. This will cause a reduction in receptor number in the brain, which will reduce the inhibition and thereby increase corticosteroid concentration. This will in turn further lower the number of receptors.

□ What is the name for the type of system that acts in the way just described?

■ A positive feedback system (Book 1, Section 7.2.2).

Under this pathological condition, the normally negative feedback system can thus exhibit positive feedback.

Evidence of elevated cortisol levels was obtained in a study on olive baboons (*Papio anubis*) living in the wild in Kenya (Sapolsky, 1990). Robert Sapolsky, from Stanford University in the USA, described the social psychology of this species in the following terms:

> With the luxury of plentiful resources and free time, the animals can devote themselves to distressing one another.

Olive baboons appear to need all their considerable advanced cognitive processing capacity for the kind of social manipulation (Book 1, Chapter 8) that forms an essential part of their day-to-day existence. Sapolsky observes that 'olive baboons occupy a social landscape of Machiavellian dimension'. ('Machiavellian' behaviour was introduced in Book 1, Section 8.3).

Dominant males have a lower background level of cortisol in the blood than subdominant males. Comparing dominant and subdominant animals, there was no difference in the sensitivity of the adrenal cortex to ACTH. Neither were there differences in the processes of breakdown of the hormone and its clearance from the blood in the urine.

□ How can the difference in cortisol levels be explained?

■ The secretion rate from the adrenal cortex must be higher in subdominants.

Sapolsky suggested that initially, on establishing a dominance hierarchy, the difference in cortisol levels would derive from the different levels of social status and control that emerge from the early experience of social interaction. That is to say, causality is in the direction (behaviour) → (hormone). However, once cortisol becomes elevated beyond a certain level, there is every reason to believe that the direction of causality can also be (hormone) → (behaviour). If the external environment is appropriate, a high cortisol level might be expected to bias the animal towards passivity and something analogous to depression.

The following section concerns the relationship between stress and the immune system (Book 2, Chapter 5), and in so doing will continue the discussion of depression.

5.5.5 Stress, the immune system and disease

Both in the scientific literature and by general opinion, stress is commonly associated with an increased risk of such infectious illnesses as influenza and colds, the implication being that stress plays some *causal* role. However, the evidence that stress is associated with such pathology in humans is fraught with controversy because of the difficulty of demonstrating that a stressor is causally related to the subsequent disease. Much of the research in this area has involved retrospective analysis of what people remember to have been taking place in their lives in the period preceding a supposedly stress-related illness.

□ Can you see any problems in interpreting the results of retrospective, memory-based studies?

■ There are two major problems. First, the fact of being ill may have sharpened the respondents' memory of stressful events occurring at about the same time. Just as many stressful events may have been occurring in periods when the respondent was well, but these have been forgotten. Second, the illness may have contributed to the respondent's subjective assessment of how stressed they were at the time, i.e. the illness may have caused the stress rather than the other way about.

In recent years, the causal sequence of (1) stress, (2) changes in the immune system, and (3) increased susceptibility to disease, has entered the popular literature in a big way. Stress and depression are indeed associated with marked deficits in the activity of the immune system, a finding that is, at first sight, consistent with the commonplace notion that people under stress are more likely to catch a cold or flu. However, the effects of stress on the immune system are complex and it is by no means clear that stress is *causally* related to a greater susceptibility to infection. The problem of disentangling the biological changes from any pathological consequences is one that is common to all forms of stress research and is amply illustrated by the effects of stress on the immune system.

A great deal of research on laboratory animals has shown that a wide range of measures of normal function in the immune system decline when stressors are applied, and recover when the stress is terminated. For example, the work of von Holst (Section 5.3) revealed that tree shrews who adopted the submissive strategy had deficits in the numbers of certain types of white cell in their circulation, compared with either the dominant or subdominant animals. Most animal studies have shown that, following periods of stress, there are alterations in the levels of antibodies in the circulation, the total number of white cells of different types, the ratio between helper T cells and suppressor T cells, the ability of cytotoxic (cell-killing) white cells to destroy infected cells close to them, the responsiveness of white cells to activating signals, and so on. All of these changes are in the direction of reducing the immune system's *theoretical* capacity to respond quickly and effectively to infection. However, evidence that stressed animals do indeed succumb to a greater number, or more prolonged episodes, of infection has been harder to come by. Animal research has been able to identify the key mechanisms by which stress exerts its inhibitory effect on the immune system, and it should come as no surprise that cortisol and the HPA system are central to the process.

Figure 5.12 (*overleaf*) revises the pathways you have met earlier by which stress is registered in the hypothalamus, then in the pituitary gland, and is reflected finally in the output of corticosteroids from the adrenal cortex. These hormones act directly to inhibit the activity of two types of white cell: the helper T cells, which you have met before (Book 2, Chapter 5) and the *macrophages*, which merit attention here. Macrophages are the largest white cell type and they have a highly specialized role in the immune response to infection. They migrate through the tissues and engulf (phagocytose) foreign bodies such as bacteria, virus particles, dead cells, debris and inorganic matter such as grit. (The progress of phagocytosis is described in Section 5.3 of Book 2.) When macrophages encounter a source of infection, they secrete a signalling molecule known as *interleukin-1*, which is an essential activating signal for all the helper T cells. In the absence of interleukin-1, helper T cells cannot secrete the wide range of signalling molecules collectively termed *lymphokines* (Book 2, Chapter 5), and hence cannot switch on and enhance virtually all the effector mechanisms involved in an immune response. Corticosteroids, therefore, are able to suppress the immune response in two ways: first, by their direct action on helper T cells and, second, by inhibiting the secretion of interleukin-1 by macrophages. Even under normal, non-stressed conditions, as levels of interleukin-1 rise during an immune response, so this is detected in the hypothalamus, and the system acts to 'damp down' the immune response by increasing the output of corticosteroids.

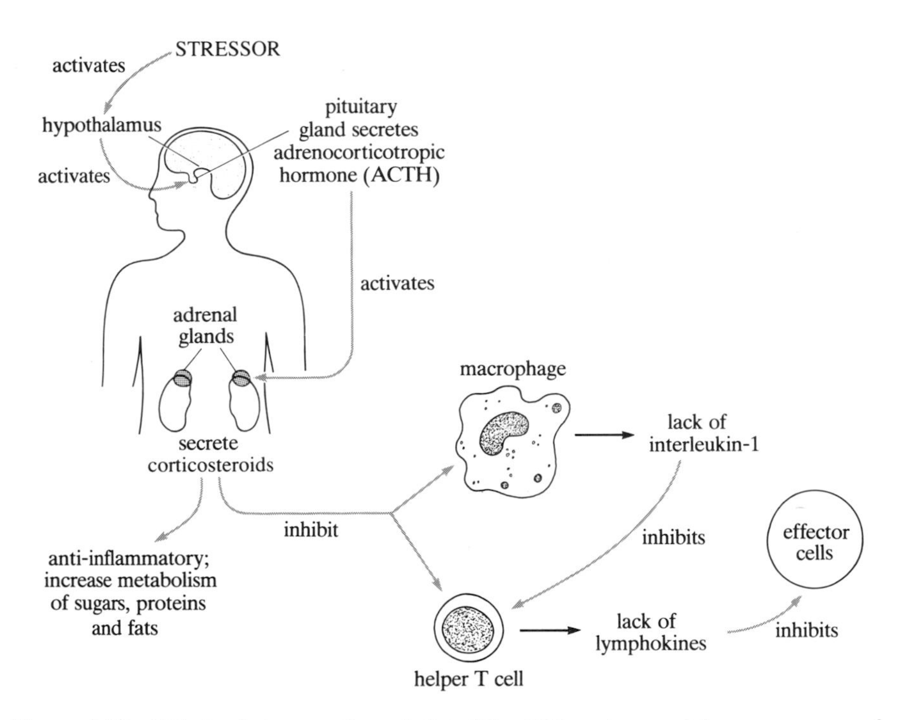

Figure 5.12 Effects of stress on the activity of the HPA system, and the consequences for the activity of white cells involved in initiating and maintaining an immune response to infection. See text for details.

However, this leaves open the functional question of why corticosteroids should suppress immune responsiveness at all? Why, following an infection, is it not simply the case that the bigger the immune response, the better? The answer seems to be that the HPA system is one amongst many elements in a web of control mechanisms by which an ongoing immune response is prevented from 'overshooting'. The function of the immune response is to destroy the source of an infection, a process which involves inflammation, but without inflicting serious damage on nearby healthy tissues. Inflammation is at least partly the result of the activity of white cells which flood into the damaged area to deal with any infectious agents that get into the body through the wound. However, inflammation is not without costs to the animal. Rather, it is associated with what is termed *bystander damage*, meaning that non-diseased tissue surrounding the target of the immune response can be harmed as a result of activity by the immune effectors. If an infection persists, despite the immune response, the bystander damage can be so severe as to be life-threatening, as is the case in tuberculosis and leprosy where most of the tissue injury is caused by the immune system. Corticosteroids are powerful anti-inflammatory agents (you may know of cortisol as 'hydrocortisone', the name it is commonly given in drug preparations), so an increase in circulating cortisol is an advantage in preventing overshoot of the immune response and its associated bystander damage. Thus, the inhibitory effect of corticosteroids on the immune response could be seen as part of a trade-off between (a) the benefit of reduced severity of inflammation following tissue damage and thereby reduced risk of bystander damage, and (b) the cost of increased likelihood of infection.

□ What other advantage does the rise in corticosteroid output confer on an animal under threat (recall Section 5.2.3)?

■ It mobilizes energy from stored sugars, fats and proteins, which is made available to fuel the 'fight-or-flight' reaction in response to the stressor.

One way of looking at the inhibitory effect of corticosteroids on the immune system under stress is to see it as a small and temporary price to pay for the tangible benefits of increasing the levels of available energy and anti-inflammatory agents in short-term stressful situations. In the natural world, stress does not usually persist for long periods; animals adopt strategies for repelling the stressor or removing themselves from the stressful situation. It is probably only in the laboratory cage or the intensive farming enclosure that animals are subjected to chronic, unavoidable stress. (Though, until more research is carried out on wild animals, it is not possible to tell whether Sapolsky's subordinate baboons, described earlier, do or do not represent an exception to what is generally found in the natural situation.) When the stress is prolonged, elevated levels of corticosteroids persist and the animal experiences long-term inhibition of its immune responsiveness. This may explain in part why intensive farming methods generally result in high drug costs to control infection in livestock, in addition to the obvious explanation that infection can spread very rapidly when animals are penned closely together.

However, an interest in stress research has been fuelled by the recognition that the natural world of humans, at least in industrialized countries, contains frequent and prolonged episodes of unavoidable stress. A driver stuck on the M25 cannot fight or flee; deadlines must be met year in, year out, by a shrinking workforce in fear of the dole queue; marital breakdown or bereavement may cause severe distress for years afterwards to those involved. In recent years, there has been a growing interest in the question of whether prolonged stress in humans results in more episodes of infectious disease. The evidence from animal studies leads to the prediction that stress must render people more vulnerable to infection, but there are serious problems when it comes to extrapolating the results of studies on laboratory animal colonies to human populations.

There are at least three reasons why experiments on the susceptibility to infection of stressed laboratory animals are inadequate models for predicting the effect of stress on infection in humans, as follows.

1. Laboratory animals are usually members of highly inbred colonies, so they are a genetically homogeneous population with very similar susceptibilities to infection when individuals are compared. Human populations are extremely varied in their genetic makeup and each individual has a different 'base-line' susceptibility to each type of infection.
2. Laboratory animals are housed in hygienic conditions and fed sterilized food pellets so, unlike humans, they are not exposed to the wide range of infectious agents in circulation in the outside world.
3. Humans exposed to a given stressor can show enormous variation in the amount of stress they actually experience—some individuals seem to thrive in the presence of stressors that demoralize others.

The immune responsiveness of human subjects under stress has been investigated in recent years, and broadly similar deficits in white cell functions to those seen in animal studies have been found. However, the evidence that these alterations in immune function lead *in practice* to an increased susceptibility to infection in humans is tantalizing but inconclusive. Most of the research on human subjects has focused on the incidence of *upper respiratory tract infections* (known as URTIs in medical jargon) because these are very common conditions, caused by a wide range of viruses and bacteria. Therefore, it can be reliably expected that a significant proportion of subjects will be exposed to some of these organisms during the study period. URTIs have the added advantage that they are easily verified by analysing throat swabs or nasal secretions, or by measuring the amount of specific antibodies to a particular bacterial or viral strain in a person's bloodstream.

A popular category of human subjects is provided by university undergraduates, who are usually assessed for the incidence of URTIs during their final exams and in equivalent mid-term periods. Another widely investigated category is elite athletes in intensive training, who appear to suffer from an unusually high rate of URTIs during the competitive season relative to other times of the year. Yet another strategy has been to get the members of a large number of different families to fill in a diary daily or weekly for several months, in which they record major stressful events. During the study period, everyone in the family is assessed at regular intervals for the presence of URTIs. At the end of the period, the pattern of stressful events recorded in the diaries is compared with the pattern of infection episodes over time. Research of this type has produced highly variable results: some studies have shown clear increases in URTIs following periods of increased stress, but others have shown no difference between stressed and non-stressed subjects.

☐ Can you suggest some of the factors that make it difficult to conclude that an increase in the rate of infection following periods of stress is actually caused by the effect of stress on the subject? (*Hint*: Think about the effect of stress on the *behaviour* of the subject.)

■ It is difficult to argue a direct causal connection between stress and infection because the stress may have produced alterations in the behaviour of the subjects, which in turn could have led to the increased infection rate.

For example, a stressed person might drink more alcohol or smoke more cigarettes, which could alter their susceptibility to infection, especially of the respiratory tract. Athletes in intensive training experience long periods of extremely deep and rapid breathing, and they might inhale more car exhaust fumes during road-running, both of which might cause damage to the membranes lining the respiratory tract and allow infectious organisms to gain a hold. It is also necessary to know the nature of the stress and whether it brought the person into more contact with infection (e.g. nursing a sick relative might be both stressful and increase exposure to infection; work-related stress might involve an increase in the number of meetings with colleagues and hence in exposure to infection).

□ Can you suggest two non-behavioural reasons why there is so much variation in the results from different studies?

■ First, the studies cannot take account of genetic differences between the people recruited for the research; variation in susceptibility to particular infections is inevitable, simply by virtue of each individual's unique genetic makeup. It could be that studies in which the stressed group experienced more infection than the controls were biased by the accidental inclusion of more biologically susceptible individuals in the stressed group. Second, the studies fail to take account of the subjects' previous history of infection.

You should recall from Book 2, Chapter 5 that once you have recovered from a particular infection, you are very much less susceptible to it if you are exposed to it again. It could be that a greater proportion of immune subjects were accidentally included in the non-stressed groups of those studies that appeared to show a correlation between stress and infection rates.

Some of these problems can be overcome by experiments in which the subject is deliberately exposed to novel infectious organisms as part of the research protocol (for example, subjects can be given a nasal spray containing a standard quantity of an uncommon strain of the cold virus). This controls the rate of exposure of each subject and allows comparison of subsequent susceptibility to that infection. An alternative strategy is to measure the recurrence rate of so-called 'cold sores' in individuals who have previously suffered from these lesions and are therefore known to be infected with the Herpes virus that causes them. Once a susceptible person has been infected with Herpes virus, they harbour it for life, but the virus particles usually remain dormant for very long periods. Stress is widely believed to be one among many factors that can trigger the re-activation of the virus, as demonstrated by sores around the mouth. The Herpes model is a particularly important one, because in some respects it mirrors the situation found in people who are infected with the human immunodeficiency virus (HIV) in its dormant phase. In the case of HIV, re-activation of the virus particles leads to acquired immune deficiency syndrome (AIDS), which has a high fatality rate, so it is vital for infected people to know whether periods of prolonged stress can trigger their progression to AIDS.

Experiments which measure susceptibility to deliberate infection or the rate of recurrence of Herpes sores in relation to periods of stress have shown a similar pattern to those described above for spontaneous infections, i.e. some studies yield convincing results and others do not. However, despite this variation in results, wherever a difference is found between stressed and non-stressed groups it is the stressed group that experiences the higher levels of infection. This strongly suggests that the effect of stress on infection rates is real, rather than an artefact of the ways in which subjects have been allocated to the stressed or non-stressed groups.

Infectious disease is only one category of pathology in which a suppressed immune system might be implicated. Rather more interest has been shown in the vexed question of whether suppressed immunity leads to an increased incidence of cancers. This hypothesis rests on the premise that everyone generates potential cancer cells throughout life, which are detected and eradicated by the immune system. The argument is that, as the effectiveness of the immune system declines with age, or with stress, so cancer cells slip through the net and become established tumours. However, the original premise has been shown to be faulty on both counts: cancer cells do not develop in most individuals until old age, and those that do arise are not detected by the immune system unless they harbour viruses or other infectious agents. In these cases, the white cells 'see' the virus rather than the cancer cell. Most of the original evidence suggesting that the immune system protects against cancers as well as infections came from experiments on laboratory animals who were treated with carcinogenic chemicals or viruses. It has been shown that these experimentally-induced cancers are unlike the spontaneous cancers that arise in humans, because they have unusual chemicals in their makeup which appear 'foreign' to the immune system. These experimental cancers can, therefore, be controlled to some extent by the immune system, and stressed animals develop tumours at a greater rate than unstressed controls.

In humans, by contrast, there is convincing evidence that the immune system is not involved in surveillance activity against the common fatal cancers, such as those affecting the respiratory, digestive, urinary and reproductive tracts, or the nervous system. For example, individuals who receive a transplanted organ are given so-called immunosuppressive drugs for a considerable period afterwards to inhibit their immune system and protect the graft from attack. Transplant patients do not suffer from more cancers of the major sites than non-suppressed people. (They do, however, have an increased risk of cancers of the immune system itself, e.g. leukaemia, but this is thought to be caused by the action of the suppressive drugs on the stem cells from which white cells are derived.) Therefore, you would not expect to find an increase in the rate of cancers in people whose immune system suffered the relatively minor levels of suppression detected after periods of stress. Indeed, long-term studies of the health of bereaved people shows a slightly elevated risk of death from a heart attack following the bereavement, but no excess deaths from cancers.

5.6 Conclusion and Summary of Chapter 5

This chapter started by looking at definitions of stress and criteria by which an animal might be said to be stressed. Four criteria of stress were discussed. Although the correspondence between these criteria is often less than perfect, it is encouraging that often a good correspondence is obtained. That is to say, the situation is stressful by the criteria that (a) it can be interpreted as thwarting the behaviour that the animal attempts to make in response to an intrusion, (b) there is an elevated and sustained input to the sympathetic or HPA systems or both, (c) the animal engages in stereotypies, and (d) the chances of a certain class of pathological change (e.g. gastric ulceration) appearing are increased.

The organizing framework for the chapter was provided by the observation that, confronted by an intrusion, an animal in the natural habitat can perform either (a) an active response (fight-or-flight), or (b) a passive response (immobility). A first attempt at a definition of stress is thus: the state that prevails when these strategies are either prevented or ineffective.

Stress was associated with a protracted effort on the part of either the SNS or the HPA system, or both. Looking at the animal's hormonal state can be a very useful way of defining conditions of stress. In spite of many disagreements, all researchers are agreed that protracted activation of the HPA system is not conducive to good animal welfare. By this hormonal criterion, chronic overcrowding is stressful, but by the same measure so is isolation for a social species.

The tree shrews studied by von Holst provide a good example of stress. The subdominants showed greatly elevated sympathetic activity, and the submissives a grossly elevated level of corticosteroids. The strategies were successful in so far as, for the most part, they served to deflect attack by the dominant. However, they failed, of course, to place the kind of distance between the two animals that would normally occur in the natural habitat.

A failure of an active coping strategy and the associated protracted activation of the SNS is associated with the risk of coronary pathology. This state was typified by von Holst's subdominant animals, which showed elevated heart rates. The failure of a passive coping strategy and the associated activation of the HPA system has important features in common with the disorder of depression. Von Holst's submissive tree shrews might serve in some ways as a useful 'animal model' of human depression.

Studies on primates similarly point to excessive stimulation of either sympathetic, or HPA systems, or both, during social conflict. Asserting dominance or challenging, being 'upwardly struggling', are associated with sympathetic activation and atherosclerosis. The more passive 'resignation' to loss of status is associated with activation of the HPA system.

The chapter discussed the concept of a coping response. For a rat, turning a wheel could be termed an active strategy of coping with the stressor when terminating the shock is contingent upon the response. Exposure to a stressor in the absence of control is associated with (a) gastric ulceration, and (b) depletion of brain noradrenalin. The latter is argued by many to be a precipitating factor in depression, as is a high blood cortisol level. Both of these physiological changes should be seen as both affecting, and affected by, psychological and behavioural factors.

Some researchers argue that performing stereotypies is at least one way of gaining some 'control' over the environment. In an otherwise impossible situation, just to behave might be better than not to behave.

If you were being lulled into a false sense of security by the development of a rather simple model of stress, Section 5.5.5 on stress and the immune system should have served to alert you to its complexity. Stress is *one* factor that can sometimes contribute to a number of types of pathology, and the evidence suggests that in some cases this relationship is mediated via the immune system. However, the qualifications have to be added that (a) other factors can often mask the relationship, and (b) there is by no means always a simple causal chain of (1) stress, (2) reduced immune response, and (3) disease. This section should have made you alert to the many pitfalls that await the unwary in examining statistics of incidence of disease and drawing conclusions from them. Even when there are clear biological differences between stressed and non-stressed subjects (e.g. in terms of heart rate, blood pressure, activity of the immune system), there is often an inconclusive correlation with pathology developed later in the two groups. Although subsequent disease rates may be higher in the stressed group than in controls, a significant proportion of stressed subjects never develop pathology, and a significant minority of those in the non-stressed group do. Lay accounts of the relationship between stress and illness reflect this paradox. Thus, on the one hand, research on risk assessment among the general public questioned about predictive factors for coronary heart disease reveals that stress appears high on the list of factors that are believed to predispose individuals to a heart attack. On the other hand, it is also widely acknowledged that some people who experience high stress remain perfectly well, while others who lead apparently unstressful lives succumb to a coronary. In addition, the focus on stress, personality and depression may divert attention away from even more important risk factors for disease. For example, the extent to which personality can be shown to contribute to the risk of a heart attack is dwarfed by the contribution of poverty, obesity and occupation.

Finally, in spite of the complexities, the rather simple model of stress developed at the start of this chapter should have served to bring some order to what is often an intimidatingly complex situation. It is hoped that, in trying now to answer the questions based upon this chapter, you will not trigger excessive activity in either of the two neurohormonal systems described!

Objectives for Chapter 5

When you have completed this chapter, you should be able to:

5.1 Define and use, or recognize, definitions and applications of each of the terms printed in **bold** in the text. (*Question 5.1*)

5.2 Describe the criteria by which a situation is defined as stressful. (*Question 5.1*)

5.3 In the context of a theory of stress, describe what is meant by the expressions 'closed loop' and 'open loop'. (*Question 5.1*)

5.4 Describe the sequence of events, involving the sympathetic nervous system, that leads to increased secretion of catecholamines. (*Questions 5.2 and 5.3*)

5.5 Describe the hypothalamic–pituitary–adrenocortical system and the sequence of events that leads to increased secretion of corticosteroids. (*Question 5.4*)

5.6 Describe von Holst's study on tree shrews in such a way that the relevance to neurohormonal systems of the terms passive strategy, active strategy, subdominant and submissive is made clear. (*Question 5.5*)

5.7 Describe attempts to fit stereotypies into a broad understanding of stress.

5.8 Describe how atherosclerosis is exacerbated by stress.

5.9 Explain the relationship between the terms 'controllability', 'coping strategy', 'closed loop' and 'open loop'. (*Question 5.6*)

5.10 Relate depression to both the sympathetic nervous system and the hypothalamic–pituitary–adrenocortical system.

5.11 Describe the ways in which stress is reflected in reduced activity in the immune system, and explain why this control loop may be advantageous in short-term stress situations.

5.12 Explain some of the methodological difficulties in demonstrating that prolonged stress leads to increased infection rates or increased cancer rates, referring to experiments on laboratory animals and on humans.

Questions for Chapter 5

Question 5.1 (*Objectives 5.1, 5.2 and 5.3*)
In the context of the model of stress, which of the following situations would be described as closed loop or open loop?

(a) In response to a ground predator, a bird takes to the air and flies away from the vicinity.

(b) An animal is threatened and it fights back. The challenger retreats.

(c) A conspecific is introduced into the cage of another. A fight ensues and the resident is beaten. The resident retreats into the corner but is still subject to regular attacks from the intruder. The resident steadily loses weight.

Question 5.2 (*Objective 5.4*)
Distinguish between the different roles that can be served by noradrenalin as: (a) a neurotransmitter in the SNS, and (b) a hormone.

Question 5.3 (*Objective 5.4*)
Injection of which of the following would be expected to increase heart rate?

(a) A noradrenalin agonist.

(b) A noradrenalin antagonist.

(c) A cholinergic agonist. (*Note* Acetylcholine is an excitatory transmitter within the PNS.)

(d) A cholinergic antagonist.

Question 5.4 (*Objective 5.5*)
Suppose an animal is observed over a period of time under constant and unstressed conditions. Its CRH, ACTH and corticosteroid levels are constant. Then some

corticosteroids are injected via an implanted surgical tube. With the help of graphs, describe the sequence of events for corticosteroid levels in the blood, CRH secretion, ACTH secretion, and corticosteroid secretion, after the injection is made.

Question 5.5 (*Objective 5.6*)
Figure 5.9 shows submissive tree shrews to have elevated adrenal glands weights. What is a possible interpretation of this?

Question 5.6 (*Objective 5.9*)
Two hypothetical SD206 students attempt a replication of the experiment on shock and gastric ulceration. First, they trained rats to turn wheels in order to earn pellets of food. Some learned very rapidly and others rather more slowly or not at all. The fast learners were allocated to the controllability condition whereas the slow learners were put in the passive yoked and unshocked conditions. It was reasoned that, since the behaviour of these latter two groups was of no consequence to the events that followed, it was best to put slower learners in these groups. Contrary to earlier findings, these students found that active rats had a greater degree of ulceration than passives or unshocked controls. What possible interpretation can be made of this?

References

von Holst, D. (1986) Vegetative and somatic components of tree shrews' behaviour, *Journal of the Autonomic Nervous System*, Suppl., pp. 657–670.

Sapolsky, R. M. (1990) Stress in the wild, *Scientific American*, **262**(1), pp. 106–113.

Further reading

Archer, J. (1979) *Animals Under Stress*, Edward Arnold.

Cohen, S. and Williamson, G. M. (1991) Stress and infectious disease in humans, *Psychological Bulletin*, **109**, pp. 5–24.

Gray, J. A. (1987) *The Psychology of Fear and Stress*, Cambridge University Press.

Stein, M., Miller, A. H. and Trestman, R. L. (1991) Depression, the immune system and health and illness, *Archives of General Psychiatry*, **48**, pp. 171–177.

Toates, F. (1987) The relevance of models of motivation and learning to animal welfare, in P. R. Wiepkema and P. W. M. van Adrichem (eds) *Biology of Stress in Farm Animals: An Integrative Approach*, Martinus Nijhoff, pp. 153–186.

Toates, F. and Jensen, P. (1991) Ethological and psychological models of motivation—towards a synthesis, in J.-A. Meyer and S. Wilson (eds) *From Animals to Animats*, The MIT Press, pp. 194–205.

EPILOGUE

The purpose of this epilogue is to take a brief overview of this book as a whole, looking for common themes and reinforcing points that arise throughout.

In the space available for the five chapters, it was impossible to do justice to the multitude of different controls that are exerted over behaviour. On reaching the end of the book, perhaps the most important thing that you should have gained is a feel for how brain and behaviour scientists approach the topic of control of behaviour. Chapters 2, 3 and 4 emphasized the caution that is needed in interpreting the results of experiments involving lesions, brain stimulation and chemical injections. On the basis of looking at the control of feeding (Chapter 2), you would now be in a good position to approach, for example, trying to understand the control of sexual behaviour. You could usefully ask what are the main sets of external factors (e.g. a mate, cues that have been paired with the appearance of a mate) and internal factors (e.g. hormone levels) that together determine sexual behaviour. You could apply an understanding of the principles of learning, looking to the background experience of the animal as a determinant of its current behaviour. If you then wished to look at the basis of the oestrous cycle, the knowledge of rhythms obtained in Chapter 3 could be applied. If you wanted to investigate what is the functional value of particular strategies of sexual behaviour, then a cost–benefit analysis of the kind described in Chapter 4 for aggression might be adapted to the case of sexual behaviour. Chapter 2 also showed how an understanding of the principles of the control of normal behaviour can illuminate an understanding of such abnormal behaviours as pressing a lever for drugs or electrical brain stimulation.

From this book, you should have formed an impression of the kind of decision-making that animals carry out. This has both causal and functional aspects. Chapter 2 described one causal aspect of such decision-making: competition between motivational systems and inhibition being exerted by one system upon another. However, the expression 'decision-making' was perhaps illustrated in its richest terms by the functional analysis of Chapter 4. Here you saw how the decision whether to fight or not is based upon assessment of the situation. The animal's perceptual processes pick up information about the opponent and then, in effect, a cost-benefit analysis is performed. Threat, escalation, attack or appeasement arise as a result of such assessment. Thus, rather than being driven passively by either external forces or inevitable and uncontrollable internal urges, behaviour is finely tuned in such a way that contributes to fitness.

Chapters 1 and 3 showed how control of behaviour is exerted both in a negative feedback fashion by deviations in body state from set points (e.g. body temperature set point) and also by rhythms. Such rhythms serve to produce activity at times when it is adaptive to be active, and sleep at times when passivity is the optimum. In some cases, feeding is aroused in such a way that nutrients are acquired before they are needed, in effect in anticipation of need.

The final chapter, on stress, described what happens to the control of behaviour when the system is 'stretched' over protracted periods. It is clear that the application of a potential stressor provokes action in terms of behaviour, hormones and the ANS that, acting in combination, can often serve to counter the stressor. However, in the situation termed 'stress' the actions are not effective in extricating the animal from the situation. The subsequent chronic activation of physiological systems (e.g. elevated heart rate, high corticosterone levels) is associated with pathology. However, whether the response should be seen simply in terms of an aberration of adaptive mechanisms under the peculiar conditions of stress is still somewhat open to discussion. There might be several functions served by these physiological changes, even under long-term stress. One such was discussed in Section 5.5.5: one of the benefits of the inhibition that elevated corticosteroid levels exert upon the immune system.

ANSWERS TO QUESTIONS

Chapter 1

Question 1.1
To make sure that the rats were sufficiently hungry to eat at the time each spaced meal was presented, which, in practice, they did.

Question 1.2
Deprive the rat of water for a period and then allow access to water only at fixed and regular intervals throughout the 24-hour period.

Question 1.3
Jill is confusing two things (1) keeping things constant and (2) defending against externally imposed disturbances. Negative feedback systems are very effective at opposing disturbances imposed upon a system. Many of the variables of the body do not move outside narrow limits. Any disturbance will be opposed. However, it is not the case that such defended variables are held absolutely constant. For instance, human body temperature shows a rhythmic fluctuation. This arises from within the temperature control system. That rats eat and drink mainly by night is explained by a rhythm within the systems that underlie feeding and drinking. It is therefore not opposed.

Ali's misconception is that the difference is in terms of the *size* of the effect, whereas it is in terms of the *origin* of the effect. External disturbances (e.g. from heat loss in a cold environment), whether large or small, will be opposed. The fluctuations he refers to in feeding and drinking are not due to external disturbances. Rather, they are due to action within the systems that control feeding and drinking.

Chapter 2

Question 2.1
(a) consummatory

(b) appetitive

(c) consummatory

(d) appetitive

Question 2.2
(a) They terminate feeding in advance of the absorption of food from the gut and cause the reversal of any internal physiological change that instigated feeding. This stops the animal ingesting too much.

(b) If they were rendered inoperative, the size of a meal following a particular stimulus to feed would be expected to increase.

Question 2.3
The simplest way would be to present it on its own for a number of times, i.e. without delivery of food following. (Another way might be to inject a dopamine-antagonist at the time of presenting the tone.)

Question 2.4
It would be expected to have little or no effect since it is involved in the termination of a meal, acting together with some other consequences of ingestion of food (though this is an extrapolation, based only on Weingarten's result).

Question 2.5
(c) Sodium depletion promotes ingestion.

(a) would make *water* on the tongue a stimulus for ingestion and sodium chloride an aversive stimulus. (b) would make sodium chloride a stimulus for an aversive reaction.

Question 2.6
(a) The hypothesis being tested concerns fear. The assumption is that the tone evokes fear because of its earlier pairing with shock. Typically, as a control, another group of rats would be exposed to the same number of shocks and presentations of the tone but they would not be paired. The analgesia evoked by the tone in these two conditions would be compared.

(b) The role of opiates could be investigated by injecting naloxone prior to the test. If the analgesia is reduced, this implicates opiates. If it is unaffected this implicates a non-opiate mediated analgesia.

Question 2.7
It is not eaten. There is a suppression of feeding during the period of egg-laying.

Question 2.8
Bill needs to qualify his assumption regarding the serotonin-antagonist. The antagonist might make the animal sick or unable to coordinate its motor activities rather than specifically targetting feeding motivation. Mary needs to take care in assuming that, because the antagonist makes them feel sick, serotonin has nothing to do with feeding. Serotonin might be implicated in feeding motivation *and* be an antagonist that makes them sick. These are not mutually exclusive. Jack's rats might have been eating at their maximum rate, as dictated by speed of ingestion or gut capacity. An agonist would be unable to increase intake any more.

Question 2.9
(a) The dopamine-antagonist could be injected and the frequency of lever-pressing compared with a period when a control injection was made. If the antagonist lowers motivation, a decline in the frequency of lever-pressing by the experimental group would be seen.

(b) Lever-pressing could be extinguished in two groups of rats. Animals in both groups could then be given one free delivery of intravenous morphine. For one group, a dopamine-antagonist could be injected prior to the priming. For the other group, a control injection could be made. Any revival of lever-pressing could then be observed. In practice, such priming tends to rearouse lever-pressing, but this effect is blocked by dopamine-antagonists.

Question 2.10

The actual experiment (Section 2.5.2) was designed to investigate the role of associative learning, i.e. to see whether a tone acquires a motivational potential by its *pairing* with the drug. To assess this, it is necessary to compare its potency with that of a tone that has not been paired with drug presentation.

Consider now the hypothetical variation on this experimental design. Rats in the control group are not exposed to the tone until the test session. (a) Suppose both experimental and control groups start lever-pressing with equal vigour. Clearly, the explanation could not be that a history of conditioning was important. It could be that presentation of the tone is enough to stress them and that this makes them resume pressing. (b) Suppose that the experimentals started pressing and the controls did not. One could not assume that it is the *association* between tone and drug that endows it with its potency. It could equally well be that simply having had a number of presentations of the tone is sufficient to give it this capacity.

Only the procedure described in Section 2.5.2 enables one to conclude that the pairing is essential.

Chapter 3

Question 3.1

The rhythm was out of phase by about 6 hours.

Question 3.2

Simply by looking at the results in this way is not enough to establish whether they are endogenous or exogenous. It is necessary to observe the animals under constant illumination (either DD or LL) to see whether the rhythm persists. The rhythms that Jane describes are known to be endogenous rather than exogenous. The light/dark cycle, the normal zeitgeber, is what entrains these endogenous rhythms.

Sean has made the mistake of taking the mean of a number of individual results and assuming that this reflects what each individual is doing. τ will vary between individuals (Section 3.4.3). Placing them under constant illumination for a while means that the rhythms can drift out of phase with each other. Taking the mean eliminates any rhythm.

Question 3.3

A circadian rhythm of activity is evident and the oestrous cycle of activity, with a period of 4–5 days, is an example of an infradian rhythm.

Question 3.4

CT0 or CT19. The frame of reference is circadian time, i.e. the subjective time of the animal as dictated by its activity profile.

Question 3.5

The result shows the phenomenon of splitting. (a) Unilateral lesions would be expected to abolish splitting, and (b) bilateral lesions would be expected to abolish the rhythm completely.

Question 3.6

This result was obtained when the animal was exposed to a light/dark cycle. The endogenous nature of the rhythm would be revealed by testing the animal during conditions of constant illumination.

Chapter 4

Question 4.1

(a) This style of definition is based on the *consequences* of behaviour.

(b) This style of definition is based on the *form* of the behaviour.

Question 4.2

The main reasons are:

(1) The developmental history of the animals may differ. For instance, males exposed to low levels of testosterone in early life are less aggressive in response to testosterone as adults.

(2) The experience of the animals may differ. Animals used to winning fights respond differently to testosterone from animals used to losing fights.

(3) The social context in which the animals were tested may differ.

Question 4.3

It is possible that the electrodes did have an effect, but not on the behaviour patterns being measured. The hypothalamus, despite its small size, is a diverse structure containing many neural systems. These 217 electrodes must have been in locations which did not influence the neural substrates involved in the behaviour patterns of interest.

Question 4.4

The relevant point here is what measure of aggression is used. In many animal studies fighting or wounds may be used as a measure of aggression. In human studies, different measures are used, such as a criminal record, or the extent of verbal abuse.

Question 4.5

A population consisting entirely of individuals showing such altruistic behaviour could always be invaded by individuals who did not behave in this way.

Question 4.6

(a) False. There are two situations in which the statement will not be true: (i) when fighting costs are high relative to the benefits to be gained from an aggressive interaction, and (ii) when individuals who escalate aggressive interactions are very common in the population, relative to individuals who do not escalate.

(b) False. Pay-off is defined as benefit *minus* cost.

(c) True. This statement is true, provided that environmental conditions do not change. For example, if resources become more scarce, natural selection may favour a shift towards strategies involving a higher propensity to escalate.

(d) False. To calculate an ESS one also has to know the relative abundance of each strategy in the population.

Question 4.7

More intense. As the value of the resource increases so the costs worth incurring increase.

Question 4.8

Roaring in red deer and croaking in toads are two good examples of assessment.

Question 4.9

The first reason is that the definition is based on the 'intention' of the aggressor. Who decides what the actual intention of the aggressor was, and is intention itself sufficient, or does some physical act have to be performed against the victim? The second reason is that the definition includes predation (e.g. snaring a rabbit for food, or catching a fish), which few ethologists would accept as aggressive behaviour.

Chapter 5

Question 5.1

(a) Closed, (b) closed, and (c) open.

Question 5.2

(a) As a neurotransmitter, noradrenalin released from a nerve terminal can influence another neuron or a particular muscle. Thereby, it can exert a local effect. (b) As a hormone, noradrenalin is released into the bloodstream, is carried in the blood, and will affect all effectors that have appropriate receptors, i.e. its effect is a 'whole body' or a global one.

Question 5.3

(a) and (d). The SNS employing noradrenalin excites the heart, whereas the parasympathetic system, employing acetylcholine, inhibits it. A cholinergic antagonist would lower the inhibition exerted upon the heart.

Question 5.4

Figure 5.13 (*overleaf*) shows this. At first, of course, the corticosteroid concentration rises sharply as a result of the injection. This increases the inhibition exerted upon the release of CRH and ACTH. A lowered release of the corticosteroid as a result of the reduction in ACTH secretion and breakdown of corticosteroids in the blood will in time bring the system back to its previous state.

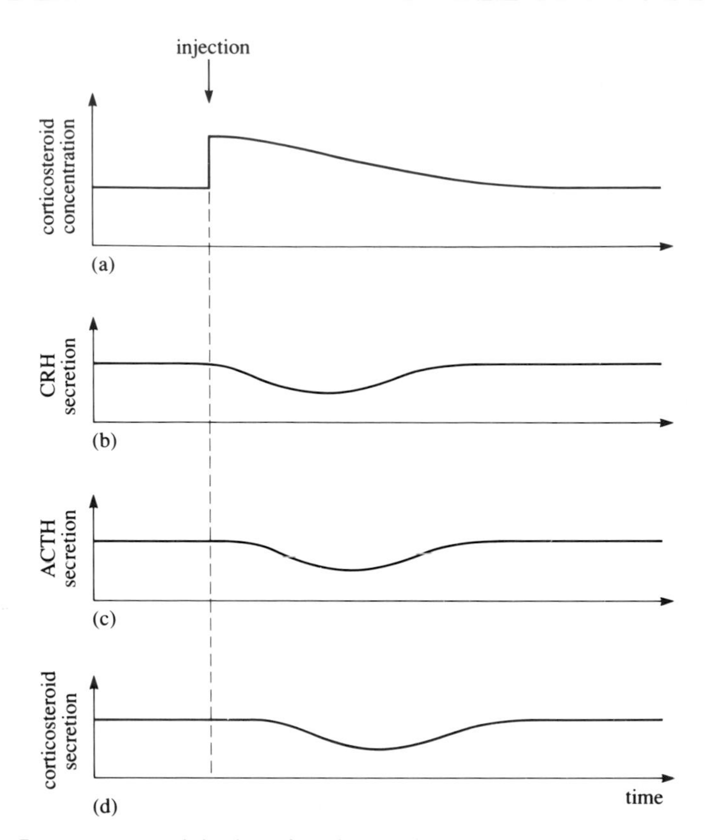

Figure 5.13 Responses to an injection of corticosteroids. (a) Corticosteroid concentration in the blood, (b) CRH secretion rate, (c) ACTH secretion rate, and (d) corticosteroid secretion rate.

Question 5.5

The blood concentration of corticosteroids in the submissive tree shrews is known to be elevated, which implies increased secretion: presumably this is reflected in increased synthesis within the adrenal cortex. Such increased synthesis would necessitate elevated levels of the precursor substances and enzymes involved in synthesis.

Question 5.6

The behaviour of the two SD206 students represents a classic howler. However, they can be forgiven through lack of experience, whereas some researchers have made just this kind of mistake with dire results. In any experiment where animals are allocated to either an experimental or a control condition, it is vital that the groups are matched as far as is reasonably possible. Consider a hypothetical but not unreasonable explanation for the SD206 students' result. In principle, it could well be that fast learners tend to be more neurotic than slow learners, and hence are more prone to gastric ulceration. Thus any effect of controllability was masked by having subjects particularly prone to ulceration in the active group.

GLOSSARY

appeasement display See *submissive posture*. (Section 4.5.2)

asymmetric contest A contest in which one of the contestants has an advantage over the other. (Section 4.8)

biological clock The timing device that underlies the rhythmicity of a biological rhythm. (Section 3.2.2)

circadian A word derived from the Latin 'circa' and 'dies', meaning 'about a day'. Specifically, it refers to endogenous biological rhythms which normally exhibit a period of about 24 hours. (Section 3.1)

closed loop The situation that prevails when a negative feedback system is working normally. In the case described in Book 5 it refers to threats instigating action that reduces the threat, e.g. places a distance between the animal and the threat. (Section 5.1.2)

controllability A situation is described as controllable if there is action available to the animal to change the situation. Possession of a coping response makes a situation controllable. (Section 5.5.2)

coping response A response that can terminate an aversive state. For example, if a shock can be terminated by pressing a lever, such a response is termed a coping response. (Section 5.1.1)

correlated asymmetry The situation that prevails in an asymmetric contest when one contestant has some *intrinsic* advantage over the other, for example it is bigger. (Section 4.8.1)

diurnal Active during the day. (Section 3.2.2)

entrain The bringing into phase with itself of a biological rhythm by a zeitgeber. For example, when a light/dark cycle has the effect upon a circadian rhythm of bringing the biological rhythm into phase with itself, the light-dark cycle is said to entrain the circadian rhythm. (Section 3.2.1)

escalation A change from threat to actual fighting. It can also refer to a change from a low-intensity threat to a higher-intensity threat. (Section 4.7)

free-running A biological rhythm is said to be free-running when it persists in the absence of the zeitgeber. (Section 3.2.2)

learned helplessness Placed in an uncontrollable situation, an animal learns that its actions do not affect the environment, e.g. nothing that it does will terminate a loud noise. When the animal is later tested in a controllable situation, it might fail to learn that it has control, i.e. it has learned to be helpless. The learning in the first phase interfered with learning in the second. (Section 5.5.3)

nocturnal Active during the night. (Section 3.2.2)

open loop A somewhat contradictory expression to refer to the situation when a feedback system is rendered inoperative as a result of opening its feedback loop. In the sense used in Book 5, it refers to what happens when actions such as fleeing or fighting are ineffective in changing the conditions that elicited them. (Section 5.1.2)

pay-off The difference between the benefits and the costs to fitness arising from a particular behaviour pattern or a particular situation, e.g. an aggressive encounter. (Section 4.7).

phase response curve A graph of the phase shift induced by light stimuli that are presented at various points in the animal's circadian rhythm plotted against circadian time. (Section 3.5.1).

priming The revival of a particular behaviour pattern performed as an operant task by presenting, under the control of the experimenter, the reward normally obtained in the task. A particular behaviour pattern that is undergoing extinction as a result of omission of reward can often be revived in strength in this way. (Section 2.4.1)

re-entrainment The process of entraining a biological rhythm to a zeitgeber different from that to which the biological rhythm was originally entrained. Thus, suppose a biological rhythm is first entrained to a zeitgeber and then the zeitgeber is changed. At first there will be a phase difference between the biological rhythm and the zeitgeber. Re-entrainment is the process of reducing this phase difference, i.e. coming back into entrainment. (Section 3.4.1)

resource holding potential The ability of an animal to hold a resource that has value for its fitness, e.g. a mate or a nest site. (Section 4.8.1)

stress-induced analgesia Analgesia (a process acting in opposition to pain) that arises as a result of the exposure of an animal to a fear-eliciting stimulus. (Section 2.3.2)

stressor A stimulus that evokes stress. (Section 5.1.2)

submissive posture (also termed **appeasement display**) Posture that has the effect of stopping a fight. Its form is often diametrically opposite to that shown in a threat display. (Section 4.5.2)

taste reactivity test A test in which substances are placed on the tongue of a rat by means of an implanted tube. The reaction of the animal to the substance (e.g. swallowing, mouth wiping) is usually monitored by video. (Section 2.2.5)

threat (or **threat display**) A gesture made by an animal in which it usually displays its potential for fighting. Weapons such as teeth are exhibited and body size is sometimes 'blown up'. (Section 4.5.2)

uncorrelated asymmetry A disparity between two animals in an asymmetric contest which is not due to their intrinsic properties but is due to chance or some external factor. For instance, by chance one animal might have arrived first at a particular spot. (Section 4.8.3)

zeitgeber An aspect of the environment to which a biological rhythm will entrain. For circadian rhythms, it is normally the light/dark cycle, but there are other possible zeitgebers such as noises and social factors. The word is derived from the German for 'time-giver' and is pronounced roughly as in 'zyte gay-burr'. (Section 3.2.2)

ACKNOWLEDGEMENTS

The book team would like to thank Dr Kent Berridge of the University of Michigan for his valuable comments on Chapter 2.

Grateful acknowledgement is made to the following sources for permission to reproduce material in this book:

FIGURES

Figures 1.1, 1.3: Reprinted by permission from Oatley, K. (1971) *Nature,* **229**, pp. 494–496, copyright © 1971 Macmillan Magazines Ltd; *Figure 1.2:* Le Magnen, J., Devos, M., Gaudilliere, J. P., Louis-Sylvestre, J. and Tallon, S. (1973) *Journal of Comparative and Physiological Psychology*, **84**, pp. 1–23, copyright © 1973 by the American Psychological Association. Adapted by permission; *Figure 2.4:* Carlson, N. R. (1977) *Physiology of Behaviour*, Allyn and Bacon; *Figure 2.5:* Reprinted with permission from Weingarten, H. P. (1984) *Physiology and Behaviour,* **32**, pp. 403–408, copyright © 1984, Pergamon Press plc; *Figures 2.8, 2.9:* Weingarten, H. P. (1984) *Appetite,* **5**, pp. 147–158, Academic Press Inc. (London) Ltd; *Figure 2.10:* Toates, F. M. and Rowland, N. E. (1987) *Feeding and Drinking*, Elsevier Science Publishers BV; *Figure 2.11:* Toates, F. M. and Oatley, K. (1972) *The Quarterly Journal of Experimental Psychology*, **24**, pp. 215–224, Academic Press (London) Ltd; *Figure 2.13:* Davis, W. J. (1979) *Trends in Neurosciences*, **2**, pp. 5–7, Elsevier Science Publishers BV; *Figure 2.15:* adapted from Wise, R. (1982) *The Behavioural and Brain Sciences,* **5**, pp. 39–87, Cambridge University Press; *Figures 3.3, 3.8:* Brady, J. (1979) *Biological Clocks*, Studies in Biology, No. 104, Edward Arnold Publishers; *Figure 3.5:* Courtesy of Dr Elena Thomas, University of California, Berkeley; *Figure 3.6:* Palmer, J. D. (1976) *An Introduction to Biological Rhythms*, Academic Press Inc.; *Figure 3.7:* Tannenbaum, G. S. and Martin, J. B. (1976) *Endocrinology*, **98**, copyright © The Endocrine Society; *Figures 3.9, 3.10:* Reprinted by permission of the publishers from *The Clocks that Time Us* by M. C. Moore-Ede, F. M. Sulzman and C. A. Fuller, Cambridge, Mass.: Harvard University Press, copyright © 1982 by the President and Fellows of Harvard College; *Figures 3.11. 3.17, 3.18, 3.23:* Courtesy of Dr Stuart Armstrong, La Trobe University, Australlia; *Figure 3.12:* adapted from Pittendrigh, C. S. and Daan, S. (1976) *Journal of Comparative Physiology*, **106**, pp. 333–355, Springer-Verlag; *Figure 3.13:* Dement, W. C. (1972) *Some Must Watch While Some Must Sleep*, W. H. Freeman and Co.; *Figure 3.14a:* Conroy, R. T. W. L. and Mills, J. N. (1970) *Human Circadian Rhythms*, Churchill Livingstone; *Figures 3.14b, c:* Fröberg, J. E. (1977) *Biological Psychology*, **5**, pp. 119–134, Elsevier Science Publishers BV; *Figure 3.15:* Aschoff, J. (1982) in *Biological Timekeeping* (ed. by J. Brady) Cambridge University Press; *Figure 3.16:* Czeisler, C. A., Weitzman, E. D., Moore-Ede, M. C., Zimmerman, J. C. and Knauer, R. S (1980) *Science*, **210**, pp. 1264–1267, copyright © 1980 by the American Association for the Advancement of Science; *Figure 3.20:* Pickard, G. E. and Turek, F. W. (1982) *Science*, **215**, pp. 1119–1121, copyright © 1982 by the American Association for the Advancement of Science; *Figure 3.22:* adapted from Schwartz, W. J., Davidsen, L. C. and Smith, C. B.

(1980) *Journal of Comparative Neurology*, copyright © Alan R. Liss Inc. 1980, reprinted by permission of Wiley-Liss, a division of John Wiley and Sons Inc.; *Figure 3.24:* adapted from Reiter, R. J. (1974) *Chronobiologia*, **1**, copyright © The International Society for Chronobiology, Milan; *Figure 3.25:* Kripke D.F., *Annals of the New York Academy of Sciences*, **453**, New York Academy of Sciences; *Figure 3.26:* Reprinted by permission from Turek, F. W. and Losee-Olson, S. (1986) *Nature*, **321**, pp. 167–168, copyright © 1986 Macmillan Magazines Ltd; *Figure 3.27:* van Reeth, O. and Turck, F. W. (1987) *American Journal of Physiology*, **253**, pp. R204–R207, The American Physiological Society; *Figure 3.28:* Armstrong, S. (1989) in *Pineal Research Reviews*, **7**, (ed by R. J. Reiter) copyright © Alan R. Liss Inc. 1989, reprinted by permission of Wiley-Liss, A Division of John Wiley and Sons, Inc.; *Figure 4.3:* adapted from Huntingford, F. A. and Turner, A. K. (1987) *Animal Conflict*, Chapman and Hall; *Figure 4.4:* Mogensen, C. J. (1987) *Progress in Psychobiology and Physiological Psychology*, **12,** Academic Press Inc.; *Figure 4.5:* Reprinted by permission from Adams, D. B. (1971) *Nature*, **232**, copyright © 1971 Macmillan Magazines Ltd; *Figure 4.7:* Schenkel, R. (1967) *American Zoologist*, **7**, pp. 319–329, The American Society of Zoologists; *Figures 4.9, 4.10:* Clutton-Brock, T. H. Albon, S. D., Gibson, R. M. and Guinness, F. E. (1979) *Animal Behaviour*, **27**, pp. 211–225, Baillière Tindall; *Figure 4.11:* Wiley, R. H. (1973) *Animal Behaviour Monographs*, **6**, pp. 87–169, Blackwell Scientific Publications Ltd; *Figure 4.12:* Davies, N. B. and Halliday, T. R. (1979) *Animal Behaviour*, **27**, pp. 1253–1267, Baillière Tindall; *Figures 4.13, 4.14, 4.15:* Reprinted by permission from Davies, N. B. and Halliday, T. R. (1978) *Nature*, **274**, pp. 683–685, copyright © 1978 Macmillan Magazines Ltd; *Figure 4.16:* Courtesy of Tim Halliday; *Figures 5.2, 5.11:* Archer, J. (1979) *Animals Under Stress*, Edward Arnold Publishers; *Figure 5.4:* Guyton, A. C. (1987) *Human Physiology and Mechanisms of Disease*, W. B. Saunders and Co.; *Figures 5.8, 5.9, 5.10:* von Holst, D. (1986) *Journal of the Autonomic Nervous Systems*, **Suppl.**, pp. 657–670, Elsevier Science Publishers BV.

INDEX

Note Entries in **bold** are key terms. Indexed information on pages indicated by *italics* is carried mainly or wholly in a figure or table.